THE IRRESPONSIBLE PURSUIT
OF PARADISE

SECOND EDITION

JIM L. BOWYER

© 2016, 2017 Jim L. Bowyer

Composition by Bookmobile Design and Digital Publisher Services, Minneapolis, MN

Levins Publishing

For individual orders go to: www.JLBowyer.com

ISBN: 978-0-9976726-1-9
Second edition

Library of Congress Control Number: 2017944402

Printed in the United States

*If we are concerned about our great appetite for materials,
it is plausible to decrease waste, to make better use of stocks available,
and to develop substitutes. But what about the appetite itself?
The major cause of the continued deterioration of the global
environment is the unsustainable pattern of consumption
and production, particularly in industrialised countries.*

—JOHN KENNETH GALBRAITH

Contents

PART 5: CHANGING COURSE

Preface

The Irresponsible Pursuit of Paradise begins with a quest to discover the secrets of Paraíso, an imaginary kingdom that is a paradise in both name and appearance. However, this book mostly dwells in the land of pure non-fiction, being an investigation of the actions of the residents of a present-day, real-life paradise as they strive to keep what they have. It is a story of unawareness, denial, hypocrisy, and a great deal of wishful thinking. And, finally, it is a story of needs for change if Paraíso is to continue to thrive.

This book is intended to call attention to several well-entrenched practices that together add up to irresponsible behavior in the present and serious trouble for future generations. On the one hand, growth is almost universally accepted as key to prosperity. Growth of populations and economies are seen as not only desirable, but essential, and this view dominates government and business thinking in the most advanced of countries. On the other hand, many of the highest consuming countries exhibit steadfast resistance to accepting responsibility for environmental impacts of their own consumption. The United States, where the economy is driven by consumer spending, is a prime example. Despite ongoing and massive consumption of basic raw materials needed to support high levels of consumption, consumer spending is foremost in thinking, with periodic reminders throughout the year of the importance of shopping to the economy. At the same time, stiff resistance to domestic extraction of raw materials has become routine, with citizens who devote time and energy to disrupting, halting, or delaying raw materials extraction viewed as heroes in some circles. And a key metric used to judge the environmental accomplishments of politicians is how much land area is placed off limits to mining, drilling, or logging. In contrast, those involved in bringing basic materials to processing centers—miners, oil field workers, and loggers—are often viewed with scorn.

Because resource extraction leads to environmental impact, actions to disrupt or halt such activity are typically justified under the banner of environmental protection. Indeed, the practice of thwarting domestic extraction of resources and associated heavy industry in the name of the environment has become something of an art form. Meanwhile, less is being done than could be to increase recycling of waste. In fact, surveys show that over 40 percent of Americans don't regularly participate in recycling programs, and that almost one-third of those aged 18–30 don't recycle at all. The cumulative result of high consumption, resistance to domestic raw materials extraction, and lack of interest in recycling on the part of many is importation of an increasing portion of raw materials used to support the economy and lifestyles of citizens—in effect an institutionalized pattern of outsourcing environmental impacts of consumption.

What this means for the U.S. is that citizens are able to enjoy the benefits of high consumption with minimal exposure to the environmental impacts of that consumption—especially in the case of metals. Moreover, these same consumers are able to bask in a feeling of environmental superiority over raw material supplying nations whose environments are not as pristine as those in the U.S. It is an ethically bankrupt position.

But there are reasons other than ethics that require self-examination of current practices. Many of the raw material exporting countries, long characterized by low per capita incomes and consumption, are now experiencing rapid economic growth and increases in consumption that accompany such growth. They are, in other words, rapidly becoming competitors for raw materials coveted by the most economically developed countries. In addition, rapid increases in basic raw material consumption worldwide have triggered concerns about environmental impacts of that consumption, as well as the potential for conflict as resource competition intensifies. These realities, coupled with changing global demographics, suggest a need to rethink the relationship of the most economically advanced countries to the rest of the world, and to fundamentally reexamine the long-term viability of consumer-based economies.

It is important to note that what follows is not about environmental gloom and doom, although a number of global realities and problems, including the daunting issue of relentless population growth, are discussed. A considerable part of the book is, in fact, dedicated to an examination of global trends across a broad spectrum. This is done with the conviction that only by realistically confronting the challenges of the present, and thinking systematically about how they all fit together, can a sound blueprint for change be drafted.

This book is written from a perspective shaped by a career in forestry and wood science that began with four summers as a forest aide, firefighter, timber cruiser, and wilderness area trail patrol in the Black Hills National Forest in Wyoming, and Arapahoe National Forest in Colorado. After several years with the Navy in the western Pacific during the Vietnam conflict, during which I gained my first exposure to tropical forests, I returned to academia to complete a doctorate in wood science and technology begun some three years earlier. The following several decades were spent as a professor in a major university, teaching and conducting research on ways to increase the efficiency of wood products manufacture and use. Inevitably my earlier interests in forest management and tropical forests resurfaced, with the result that my research and teaching increasingly dealt with implications of wood use on forest conditions and forest health. This work included extensive travel to temperate, tropical, and boreal forests around the world, as well as a number of years of tracking national and global trends in consumption of wood and other raw materials.

For the last fifteen years of my academic career, my teaching and research became focused on environmental topics including environmental life cycle assessment. During that period I also formally surveyed the environmental knowledge and attitudes of thousands of people across the U.S. and Canada. In the 36+ years spent in academia I had the opportunity to interact with students in classroom settings, and with hundreds of audiences and literally tens of thousands of people representing a broad cross-section of society at large. Since that time I've worked as a consultant on a variety of environmentally-related problems and issues with both the public and private sector. These experiences have provided considerable insight into what and how people think about the environment and environmental issues.

Given the breadth of topics covered in this book, any one of which could alone be the subject of a tome, it was tempting to write a volume of a thousand pages or more, examining every aspect of every issue. Instead, I decided to be as succinct as possible in capturing the essence of the central issues discussed, and essential relationships between these issues and myriad global realities. Note that several topics recognized as vitally important—food and food systems, fresh water supply, and climate change—are nonetheless referenced only briefly. Questions are posed at the end of the book for which direct answers are not provided. The hope is that these will help to stimulate thinking, introspection, and discussion with others.

I am indebted to a number of people who have read through various drafts of the manuscript and provided valuable feedback and suggestions for change. Special thanks to Ruth Bowyer; Jeff Bowyer, environmental consultant to the Asian Development Bank; Dr. Jeff Howe and Dr. Steve Bratkovich of Dovetail Partners; Dr. David Morris, Co-Founder and Vice-President, Institute for Local Self-Reliance, Washington, D.C.; Jennifer O'Connor, President, Athena Sustainable Materials Institute, Ottawa, Ontario; Dr. John Tilton, Research Professor and University Professor Emeritus, Division of Economics and Business, Colorado School of Mines, Boulder; Dr. David Foster, Department of Organismic and Evolutionary Biology, Harvard University; Dr. John Helms, Professor Emeritus, Department of Environmental Science, Policy and Management, University of California, Berkeley; and Dr. Robert Malmsheimer, Professor, College of Environmental Science and Forestry (ESF), State University of New York, Syracuse for their extensive review through several revisions. Special thanks to Richard Levins of Levins Press for his insightful suggestions through several drafts, and for his considerable assistance along the way, and to Connie Kuhnz of Bookmobile for her work in bringing the draft to finished form. Thanks also to my amazing colleagues at Dovetail Partners— Katie Fernholz, Dr. Jeff Howe, Dr. Steve Bratkovich, Dr. Ed Pepke, Harry Groot, Matt Frank, Gloria Erickson, and Dr. Sarah Stai for the ongoing and spirited engagement in examining and debating a wide range of environmental topics.

JIM L. BOWYER

North Oaks, Minnesota

THE IRRESPONSIBLE PURSUIT OF PARADISE

SECOND EDITION

Paraíso!

Roaming the pristine grounds of castles high in the hills,
I wished that time would stand still.

—ELISA MALA

As Daniel left the border crossing behind and began the long journey home, he knew he would come to dread the long hours in the saddle before reaching his destination. Today, however, he was in good spirits. He had, after all, just spent over a month traveling the width and breadth of a Kingdom that he had wanted to visit since he was a young boy. He also had an answer to the riddle posed to him by Lord Jardel. Daniel had gone to his lordship to ask for a letter of introduction that would allow him to travel to Paraíso, and also for modest funds to finance his journey. To his surprise he was, with a bit of hesitation, granted both, but with one stipulation: Daniel was charged with the task of finding out how the Kingdom of Paraíso managed to provide its subjects with the highest living standard in all the world, while also maintaining a reputation as the most beautiful kingdom of all.

Daniel had first heard tales of the fabulous kingdom as a child, and as the years passed, his interest had only grown. What intrigued and inspired him were reports of unmatched wealth, manifested in the form of citizens who had more in the way of worldly goods than even the nobility of most other kingdoms. Even average Paraísoans were said to own spacious cottages, to have two and sometimes three horses, and to have enough gold to travel often to all parts of the kingdom and beyond. Reports indicated that some commoners even had two cottages! To Daniel, such things were simply unimaginable; in his village cottages were smaller than small, few families were lucky enough to own a donkey, much less a horse, and almost all the energy of villagers was expended in obtaining enough food and cooking fuel for the coming day. Travel outside the immediate area of the village was simply unthinkable, and in any case was something reserved for only the knights of the realm.

He thought back to his first days within Paraíso. He had been astonished at the sights and sounds. Wide streets and thoroughfares and even alleyways of continuous stone; village after village of sometimes two and three story cottages, each with an attached stable, each with manicured grounds befitting the finest town squares in his memory; everywhere fancy carriages, coaches, and carts, many quite large and often pulled by heavy, straw-consuming draft animals; markets and bazaars beyond anything he had ever seen or dreamed of—row after row after row of meats, vegetables, fruits, breads, soups, and spices within colorful and lavishly appointed tents and what appeared to be common people carrying it all away by the sack-full, even while others brought forth still more food to be sold; merchants offering expensive shoes, clothing, jewelry, perfumes, and goods of all kinds from kingdoms around the world in massive cathedral-like emporiums, again with throngs of commoners everywhere examining, trying-on, and buying, buying, buying; scores of taverns and eating halls, some with substantial bowls and plates that were simply thrown away at the end of the feast. Incredible! Unbelievable!

As incredible as the bountiful trappings of untold wealth were, so too was the setting. Whereas his own village operated under a continuous pall of dust and smoke from the nearby mines and metal smelters and was crisscrossed by several creeks the color of caramel, both the air and the water in Paraíso were amazingly clear, the water reflecting the brilliant blue of smoke-free skies on sunny days.

After several weeks of travel and similar scenes in town after town, village after village, Daniel's amazement had turned to serious contemplation and the words of Lord Jardel had echoed in his ears. Daniel's request for travel funds had been met with a long uncomfortable silence such that he had been tempted to humbly withdraw his request and depart in haste, but just as such thoughts were taking shape Lord Jardel began talking as if to no one in particular. "I don't understand," he had said, "how they manage to take so much from the Earth, and yet live in a kingdom that has from north to south and east to west the appearance of heaven itself." Abruptly turning to Daniel, Jardel had smiled, said "your requests for a letter and travel funds will be granted," adding "but in return, I want you to learn what you can about the riddle of riches and heaven. Do that, and I will be in *your* debt."

Now as his horse broke into a trot at the sight of the river ahead, Daniel was confident that he knew the answer. His first glimmer of understanding had come one sun-splashed afternoon as he stood at the edge of a clearing and watched an army of woodworkers erecting cottages made of heavy wooden timbers covered with wooden staves. Thinking back to the park-like forests he had traversed in previous days it dawned on him that on only two occasions had he encountered woodsmen cutting trees or hauling timbers, and in fact had seen very few locations from which trees might have been removed in recent years. From that day forward he had paid a good deal of attention to the countryside, asking many questions along the way. In a village where artisan after artisan displayed stunning wares of pewter and silver he had asked in what direction he should go to visit the mines. "To the Kingdoms of Quel or Chemere, I would suppose," was the reply. Beside a large battlement where soot-blackened peasants were shoveling coal into a basement window he had asked the bored-looking overseer where the coal had been unearthed; "to the South—in the Kingdom of Legbot, from whence most of these poor helpers have come" the officer had grunted. And when he came upon a man leading a pair of oxen that were pulling a cart loaded with timbers he asked whence they had come. "Why from the Kingdom of Cander from whence most of our timbers come," the man had said. Paraíso, it appeared, was simply using its incredible wealth to buy vast quantities of resources from others, inflicting in the process insult to the forests, hills, and fields of kingdoms such as his own; not such a complicated riddle after all!

In the days that followed, Daniel completed in his mind the report that he would deliver to Lord Jardel. He would tell his lordship in appropriate detail where he had gone, what he had seen, about the army of carpenters, what he had learned about sources of metals, fuel, and timbers, and then his conclusions regarding the riddle. He would tell of the large numbers of peasants crossing the border from Legbot, despite the best efforts of the King's army to stop them. He would mention his visit to the royal library and his studies of the royal ledgers. Finally, he would add the observation that what Paraíso was doing couldn't possibly last much longer in the big scheme of things, since despite the kingdom's incredible wealth the net outflow of gold would surely drain the royal treasury over time.

PART 1

THE PROBLEM

The situation described in the mythical kingdom of Paraíso unfortunately has a modern day analogue. Several of the world's most economically developed countries, and perhaps especially the United States, exhibit similar tendencies.

The problem stems, in part, from widely-held misconceptions, and is perpetuated by a general unwillingness to reconsider long-established social norms, or to confront challenging and uncomfortable issues that might lead to change. The result is inaction even as the global landscape is shifting rapidly.

CHAPTER 1

A Conundrum

"We start with Extraction.
Which is a fancy word for natural resources exploitation,
which is a fancy word for trashing the planet."

—*THE STORY OF STUFF*

The words *sustainable* and *sustainability* are sometimes used when considering the long-term effects of human activity on the global environment. Many writers have sought to define the term "sustainable." A favored definition is that from the groundbreaking 1987 report of the World Commission on Environment and Development (the Brundtland Commission report):

> *Sustainability means "meeting the needs of the present without compromising the ability of future generations to meet their own needs."*

Fundamentally, then, the question is whether current human activity is altering the Earth's biosphere and natural systems as to so degrade them over time that they cannot meet the needs of succeeding generations. Stated differently, can we continue on more or less our current path for a long time—say, hundreds or even thousands of years—to come?

It is an interesting exercise to contemplate what kinds of human activity are and are not sustainable. Is, for example, the current U.S. population growth rate of less than one percent annually something that can be sustained over the long term? What about the increase in consumption that will inevitably accompany population growth?

On a related note, is it likely or even possible for the human population to grow by 3 to 4 billion while simultaneously maintaining all of the world's biodiversity; all of the world's remaining indigenous cultures, hunting grounds, and sacred areas; all of the world's current expanse of tropical forests? And if it isn't, is that sustainable?

Maybe. Possibly. How, then, do the answers to these questions change if the global economy grows at a far more rapid rate than population, and if per capita consumption continues to rise in developed and developing nations alike? Can increasing demands and expectations be accommodated? Will natural landscapes and associated life forms and natural systems be obliterated or substantially diminished in the process? Do we face a choice between the lifestyles we want and a sustainable future? Can we find a way to accomplish both?

The answer to these and related questions will be found only by working within a broad, fact-based context while being mindful of the complexity of the many challenges and their interrelationships. In addition, if we are to ensure that the world has an adequate and sustainable supply of "stuff," we need to employ a clear understanding of where that "stuff" comes from and what the tradeoffs are in obtaining it.

Solving Complex Problems

It is widely recognized today that there are three essential pillars of sustainability—ecology, economy, and society—and that achievement of sustainability requires attention to all three pillars simultaneously. A society cannot, for example, create a robust economy over the long term in the absence of a sustainable environment or a social fabric that addresses the essential needs of its citizenry. Similarly, a society cannot achieve a sustainable environment in the absence of a sustainable economy and/or social structure that can provide the financing and support base for environmental action. Nor can a sustainable society be built without also building a sustainable economy and maintaining a sustainable, life-supporting environment. These realities provide the context within which a political or community leader, an environmentalist, or anyone else seeking to address an environmental issue must operate.

The environmental problems faced by society today are principally multidimensional, complex problems that cannot be easily solved. The easy-to-solve problems have for the most part been dealt with long ago. If we are to have any hope of finding lasting solutions to current and future environmental problems at least four elements must be part of environmental planning and decision-making processes:

- Planning must be global, systemic.
- Information on which decisions are based must be scientifically verifiable.
- Where scientific knowledge is lacking, carefully considered assumptions must be employed. Such assumptions must be realistic, and as complete as possible.
- Decision-making must be based on rational thinking.

It is abundantly clear that actions taken within one region or state or nation can impact the environment of other regions, states, or nations. Obvious examples are water and air pollution, including greenhouse gas emissions, which move freely across boundaries. Less obvious, but no less significant, are actions that may shift environmental impacts of consumption from one region to another. Because of these realities, careful consideration must be given to the potential impacts of local actions on distant regions.

Of even more importance than consideration of impacts beyond regional boundaries is systems-oriented planning. The entire world is composed of a vast number of systems—ecosystems, political systems, economic systems. Such systems, large and small, operate continually and often interlock with one another.

What is vitally important to understand in planning is that when a change occurs within a system (such as an economy or ecosystem), that change almost invariably triggers other changes in the system. And, because economic, social, and environmental systems are inextricably linked, changes within one system can also lead to changes in another. Moreover, such changes are often as significant, or even more so, as the change that triggered them. The implications for planning are obvious: unless planning processes successfully anticipate changes that might result from an action or policy shift, unanticipated changes can easily negate or even overwhelm the desired outcome.

As to assumptions, it is almost always necessary to make them. Seldom do we have all the information available that is needed in planning, nor are all the linkages within and among systems exactly known. In these situations, then, it becomes necessary to estimate or assume relationships and reactions. But estimates cannot be made casually. Just as bad data gives meaningless results, bad assumptions do as well. Each assumption must be carefully considered and reconsidered. And, if a given assumption does not stand up to close scrutiny, then planners must discard it and replace it with a better one.

It is probably unnecessary to point out that sound decision-making requires rational thinking. Nonetheless, there is today a remarkable lack of rationality, realism, and global and systemic thinking applied to environmental planning, and perhaps especially within the United States. This situation must change if progress is to be made in solving the most serious environmental problems facing the world.

To make matters more complex, we live in a dynamic world. Things are constantly changing over which we have little control. Even the problems themselves are moving targets. It appears that the world population will increase by two to four billion people within this century. The U.S. population will nearly double within the same time frame. And as domestic and global economies grow even more rapidly than the expansion of population, the combined effect will be a need for more space, food, housing, clothing, energy, durable and non-durable goods, and raw materials of all kinds. Because of these changes, environmental concerns will be magnified even further. Sustainability questions will loom even larger.

Those currently tasked with resolving these issues—land managers, government agencies, and resource-using industries in the United States and around the world—are increasingly expected to protect and preserve natural landscapes and associated values, including biodiversity and indigenous peoples, while at the same time fulfilling the world's need for basic raw materials. They are expected to do this, moreover, by a public that is almost totally disinclined to realistically face up to the daunting issues of population growth and rising consumption.

In the absence of a new approach to environmental planning, disagreements over what to do are likely to become even deeper and conflicts sharper. It is not hard to imagine a future in which interest groups are more prevalent, larger, better financed, and even less willing to compromise than today, yet just as reluctant as society at large to take on the major factors underlying sustainability concerns. In such a future, the ability of future generations to meet their own needs will almost certainly be compromised, despite the rhetoric of today that suggests intentions to do otherwise.

Stuff—It Has to Come from Somewhere

Few things stir up more emotion than the extraction of industrial raw materials. Whether drilling for oil, mining for coal or iron ore, or harvesting timber, the activity and those involved are often viewed with disdain. An example of the extent to which extraction is detested is provided in *The Story of Stuff,* described as "one of the most-watched environmental-themed online movies of all time." With over 40 million views as of late-2015, the apparent success of this production is epic. While the video does call attention to the important topic of environmental impacts of consumption, it unfortunately does so through frequent use of over-generalization, sensational language, and factual misinterpretation.

That the narrative is hard-hitting is undeniable, as indicated by this short excerpt:

> *We begin with Extraction. Which is a fancy word for natural resources exploitation, which is a fancy word for trashing the planet.*
> *What this looks like is that we chop down the trees, blow up the mountains to get the metals inside, we use up all the water, and we wipe out all the animals . . .*

. . . Now I know this can be hard to hear, but it's the truth and we've got to deal with it.

With this as a starting point, think for a moment about all of the things that you have consumed or made use of in the past week. Then, make a written or mental list (using the list below, if needed, to stimulate your thinking) of any of these things that did not have its origin in mining (including drilling), forest harvesting, or agriculture (including fishing).

Air conditioner	Chocolate	Lampshade
Alarm clock	Coca-Cola	Lawn mower
Apple	Coffee	Light bulb
Aspirin	Comb	Lipstick
Automobile	Crayon	Magazine
Backpack	Computer	Mattress
Bandage	Clothing	Milk
Baseball bat	Dinner plate	Money
Basket	Drapes	Napkin
Bathtub	Driveway	Newspaper
Bed sheets	Energy	Paint
Bicycle	Envelope	Shoes
Boat	Eraser	Park bench
Books	Eyeglasses	Pencil
Breakfast cereal	Fork	Perfume
Broom	Frying pan	Picture window
Cabinets	Furniture	Plant fertilizer
Calendar	Garden tools	Ping pong ball
Camera	Hairspray	Postage stamp
Camping gear	Hammer	Purse
Candle	Hanger	Radio
Candy bar	Hand lotion	Razor
Cardboard box	House	Refrigerator
Carpet	Ice cream	Salt
Cell phone	Keys	Scissors

Shampoo	Super glue	Toothpaste
Shopping bag	Tackle box	Towel
Soap	Television	Trash can
Soccer ball	Tennis racket	Umbrella
Steak	Tires	Watch
Suitcase	Tissue	Wrapping paper

It is highly likely that you don't have anything on your list. You may have listed water, but chances are good that before that water got to you it went through a treatment plant in which various chemicals and filtration processes were applied—all of which use energy and consume mineral resources.

The fact is that virtually all of the products each one of us uses every day have their origins in mining, forest harvesting, or agriculture. Like it or not, we all depend heavily on the blood, sweat, and ingenuity of miners, drilling crews, loggers, farmers, and commercial fishermen and women. Likewise, we are heavily dependent on what these people do—mining, drilling, logging, farming, and fishing.

That we are so very dependent on raw materials extraction and the processing that follows makes societal abhorrence to such activity a bit of a curiosity, and especially within high-consuming societies. Perhaps we have largely forgotten where things come from, or the inputs, risks, and effort required to produce a bushel of wheat, a pound of steak, a ton of steel, or a load of lumber. Or, perhaps we have for so long been so successful in avoiding exposure to the environmental impacts of our consumption that we find surprising and even shocking what those impacts really are.

The process of gathering and processing raw materials—including the collection and processing of materials for recycling—results in environmental impact. This is true of all categories of materials, though for some the impacts are far greater than for others. And, extraction of virgin raw materials generally leads to greater impacts than recovery of materials for recycling, a reality that may explain why those involved in recycling are regarded as heroes, while those associated with extraction are sometimes viewed with scorn.

Stiff resistance to domestic extraction of basic raw materials has become routine, as has activity to develop ever more restrictive environmental laws and regulations that impact raw material extraction and
processing. Though opposition of individuals and communities to heavy
industrial activity, potential changes to beloved landscapes, increased
transportation, and other aspects of extraction is completely understand-
able, and completely justified in some cases, a significant problem arises
when such opposition becomes pervasive—especially within a high-
consuming society such as the United States. Raw materials must come
from somewhere. So when raw material extraction or heavy industry is
blocked in one locality or region, then those activities and all associated
impacts are simply shifted to some other locality or region.

Actions to thwart extraction are sometimes justified using references
to "protecting natural heritage" or "natural capital." Never, however, is
consideration given to who else's heritage is likely to be significantly
modified or forever despoiled. Nor is the possibility of finding ways to
reduce consumption of the material in question ever seriously considered.

As raw materials imports have increased, so too has recovery of discarded materials for recycling. However, it is all-too-clear that society is
doing less than it could in this regard, with many citizens choosing to not
participate in recycling activity.

Due, in part, to the combination of high consumption, aversion to environmental and esthetic impacts of domestic raw materials extraction,
and less than optimal recycling practices, the United States is a net importer of all kinds of industrial raw materials, and on a truly massive
scale. Massive means 40–50 percent and more of domestic needs of most
metals, and 100 percent of many categorized as relatively scarce and
high-impact metals. What this means is that U.S. consumers are able to
enjoy the benefits of high consumption with near total freedom from exposure to the environmental impacts of that consumption—especially in
the case of metals. Moreover, these same consumers are able to bask in a
feeling of environmental superiority over raw material supplying nations
whose environments are not as pristine as those in the U.S. This is a world
view totally removed from reality.

However, change—fundamental change—is underway. What is coming in the relatively near term is nothing less than a reordering of the world hierarchy. Nations long thought of as woefully underdeveloped, if thought of at all, will increasingly command attention. Many of these nations will achieve (or are already well along the path to achieving) marked and sustained economic growth. Economic growth, in turn, will be accompanied by increasing domestic consumption of material goods (and raw materials), as well as increasing ability to compete for raw materials globally. At the same time, nations long recognized as economic powerhouses will experience erosion of their once-dominant positions. Even the balance of military power is likely to change.

China provides a stunning example of rapid change. China's ascendance as a global economic power is all too evident to the weekend shopper. Whether seeking clothing or furniture, hardware or home goods the "Made in China" label is ubiquitous. To economists, the People's Republic is an economic marvel, having chalked up a 25-fold increase in Gross Domestic Product (GDP) per capita since 1980. In the process it has risen from the world's 134th largest economy (1980), to 1st or 2nd today, depending upon the measure used. To the business world, China is a resource consuming juggernaut, alone accounting for consumption of 40 percent of the world's base metals in 2010. The shift in fortunes is almost unimaginable.

But as awe-inspiring as these developments are, China's accomplishments to date are in many ways merely the harbinger of much greater change to come. China is not the only nation in which change is rapid, nor is economic gain limited to countries now gaining the rapt attention of global financial markets—India, Brazil, Indonesia, Vietnam, and Poland. The economic genie is out of the bottle all over the world, creating a new reality for all nations, and profound implications for the select club of the most economically developed nations that for the past six or seven decades have pretty much set the world agenda. How these nations, including the United States, interact with the rest of the world, maintain their economies, obtain and use raw materials, and protect their environments will undergo fundamental change.

All of this leads to what appears to be a conundrum, expressed below in the form of three questions for citizens of the United States and other economically developed nations:

- If the current strategy is maintained without change, despite the moral shortcomings of that strategy, is our country likely to be able to continue to successfully compete for the raw materials needed to support the economy and current lifestyles? And if not, what then?
- If consumption levels are voluntarily reduced, won't that ruin the domestic economy, making environmental protection less affordable and therefore less possible?
- If consumption is not reduced, but steps are taken to bear more of the environmental consequences of that consumption, won't that ruin the domestic environment outright?

References

The Story of Stuff Project. 2008. *The Story of Stuff.*

World Commission on Environment and Development. 1987. *Our Common Future: A Report to the UN General Assembly.* Oxford University Press.

CHAPTER 2

Sourcing Globally, Protecting Locally

*". . . we are between two eras, i.e., we are in a transition
from an old era, an industrial society, to a new era,
an information society."*

—JOHN NAISBITT

On a daily basis, and at all levels of government, environmentally-inspired decisions are made in the U.S. and other developed nations that are designed to prevent, halt, or markedly restrict resource extraction and primary processing. In region after region, the environmental impacts of mining, smelting, drilling, timber harvesting, and waste disposal are deemed unacceptable. The message conveyed by activist groups and quite often mimicked by the general population is strikingly similar— Not in *My* Backyard!

While to much of American society resource extraction is "out," consumption is most definitely "in." Indeed, the U.S. government periodically appeals to the citizenry to support the economy by spending and consuming more. Television news programs during the holiday season often lead with reports on levels of consumer spending during the previous 24 hours. Although falling sharply during the recession years of 2007–2010, consumption is again rising as the economy rebounds.

It is apparent that increased consumption requires increasing quantities of resources to support that consumption. The implications of public resistance to domestic production of basic materials in the face of increasing consumption are also apparent: local actions to halt or prevent resource extraction are not stopping these activities at all. The effect is to simply shift them to some other location—to someone else's backyard.

More often than not, someone else's backyard is located in a developing country. In the mid-1970s following the first Earth Day, the slogan "Think Globally, Act Locally" was widely recognized across the United States. Appearing in the form of bumper stickers and billboards, the message resonated with much of society and appeared to signal a new level of environmental awareness. Subsequently, however, the idea of global thinking simply disappeared, leaving local action as the centerpiece of environmental thinking.

Opposition to domestic gathering and processing of raw materials was given a bit of momentum in the early 1980s with the publishing of John Naisbitt's best seller *Megatrends*. Naisbitt presented the view that the U.S. had passed from an industrial society to an information society and that jobs would increasingly be focused on the business of developing and conveying information to the rest of the world. This was all the encouragement that some needed to launch a vigorous attack on raw material extraction.

A look at what has happened since the early 1970s makes Naisbitt's observations difficult to dispute. The United States has moved into the computer age, and the impact on the economy, jobs and people's lives has been tremendous. However, one thing that Naisbitt neglected to point out is that a high-consuming nation such as the U.S. tends to consume massive quantities of raw materials whether it is an "industrial-oriented society" or not. Further, the fact that the U.S. had changed from a raw material-exporting nation in the 1950s to a net importer of raw materials on an ever-greater scale by the 1970s and '80's and into the early part of the 21th century, appears to have escaped attention altogether.

State of Denial

As a consuming region, California is among the world's elite, with its citizens enjoying one of the highest standards of living anywhere in the world. As its name suggests, it is truly the "Golden State." However, California has also been recognized by another, less favorable nickname. Pulitzer Prize winning journalist Tom Knudson spotlighted the large and growing international environmental impact of U.S. consumption with the series *California—the State of Denial.*

Knudson noted that California is, at the same time, the nation's most prodigious consumer and net importer of petroleum, wood products, and sand and gravel, among other resources. The net import situation exists, moreover, despite the relative abundance of all of these resources within that state's geographic boundaries. Knudson reported that these realities persisted against a backdrop of ongoing, vigorous initiatives to limit local production of resources even as local consumption of all raw materials was steadily rising. Moreover, there was (and is) no evidence of significant actions of any kind to limit or even question local consumption of resources.

Findings were, in short, that one effect of prohibition of new drilling activity within international waters along the California coast was to stimulate oil exploration and drilling in Ecuador, a country with few environmental regulations, with devastating environmental and social consequences to the Ecuadorian region of the Amazon. Timber harvests were also shifted elsewhere—in this case largely to Canada, partly because of environmentally-inspired actions within the Golden State to sharply curtail local logging. Similarly, citizen opposition to local sand and gravel operations shifted a significant portion of that activity to Mexico and British Columbia, resulting in significant impacts to rivers and streams, disruption of local communities, and adverse impacts on public health. This activity also triggered increased emissions and other environmental impacts linked to long-distance transportation.

Although California was the subject of Knudson's investigation, a similar situation regarding consumption and net imports characterizes most other U.S. states and many of the developed nations around the world. In short, similar findings regarding petroleum, wood products, metallic minerals, or sand and gravel could have come from almost anywhere in the developed world.

Timber

One of Knudson's most interesting revelations focused on California's timber supplies. He observed that in the mid-1950s California was self-sufficient in wood, but that by early in the 21st century, driven by aggressive efforts to protect its environment, the state imported 80 percent of what it used. At that point, forest harvest levels within the state were less than 30 percent of what they had been a half-century earlier, despite the reality that consumption of wood in California was rising steadily and the fact that annual growth in California's forests was more than double the annual rate of removals and mortality. In 2013, harvest levels remained about the same as in 2002, while net annual growth was estimated to be 4.5 times greater than annual removals. Knudson noted that the dramatic shift, from self-sufficient to massive net importer, the result of environmental lawsuits, public opinion, and increasingly strict regulations, had the effect of simply shifting the environmental impacts to Canada. In fact, logging to supply wood for California consumption not only shifted to Canada, but also to other regions. Foreign imports of wood (primarily from Canada) increased by over 40 percent from the mid-1990s through 2008, while imports from other states increased by 90 percent during that period.

As with other environmentally-inspired initiatives, there is no record of any discussion in the course of court deliberations, legislative hearings, or development of state agency regulations regarding where wood to supply California's consumption might come from if not from within the borders of the state. Nor was any thought given to ways in which consumption of wood might be reduced.

Actions to "protect" California's forests had at least two unintended consequences:

- Aggressive curtailment of harvests in California forests contributed to increases in the volume of woody biomass in the states' forests that had been building up over a number of decades, a process that continues today. Biomass stocks are currently estimated to be far above historic levels, a situation that greatly increases the odds of disease and insect infestation and catastrophic fire events.
- Ever-intensifying forest practice regulations, especially as a result of rule amendments in the early 1990s, resulted in cost increases in developing timber harvest plans of up to 1,000 percent (an average of $30,000 by 2005 as compared to $2,500 30 years earlier). Consequently, many California timberland owners opted to sell their land for higher

returns, frequently resulting in conversion of forested land to housing. In the words of a team of investigators that examined forest trends, "California's increasingly strict environmental regulations of forestland are, in many cases, having precisely the opposite effect from that which was intended."

Sand and Gravel

Sand, gravel, and crushed stone together are defined as aggregate. Mixtures of these materials, and cement—the active ingredient in concrete—are used in building roadbeds and in concrete or asphalt highway surfaces. They also are essential ingredients in concrete structures of all kinds. California alone uses 200–240 million tons of this material annually.

Sand and gravel reserves in California in 2003 were estimated at about 81 billion tons—enough to supply the state's needs at 2003 rates of extraction for 350 years. Yet, California became a net importer of both materials at about the turn of the 21st century because of, according to Knudson, higher priority placed on building houses than on preserving access to highly concentrated reserves, and concerns about potential impacts on the environment and recreational areas.

In this case, impacts of mining were shifted to Canada and Mexico, which together supplied abut about 1.5–2 million tons a year to California in 2000. In 2003 Mexican authorities reacted to increased shipments, halting a rail shipment of sand destined for San Diego. One such official was quoted in the *Los Angeles Times* as saying that if sand continued to flow to the U.S. at the current rate there would not be any more sand in Baja, California. Knudson related comments of a number of Baja residents who cited ongoing damage to rivers and surrounding landscapes as a result of increased aggregate extraction for shipment to California, and growing resentment of what was seen as a practice of protecting the U.S. environment at the expense of distant environments, communities, and lifestyles. Similar sentiments were expressed north of the border in British Columbia. Also related were the comments of a senior geologist with the California Department of Conservation, who noted that the permitting process for sand and gravel mining had become progressively more expensive and time-consuming. The effect was to drive companies out of business and to decrease the number of mines. But consumption of sand and gravel did not decrease. As of early 2016 the flow of sand and gravel to California continued. Though both Mexico and Canada continue to supply material, Canada has become the dominant supplier. A 2013 report "California's Looming Sand-and-Gravel Crunch" suggests that demand for greater imports may be on the horizon.

A Broader Problem

As indicated previously, the phenomenon described by Knudson is not limited to California. Variations of the same story are replayed again and again across the U.S. and other developed nations. In the name of helping the less fortunate and protecting the environment, the practice of thwarting domestic extraction of resources and associated heavy industry has become common.

The global impacts are not always immediately obvious. An example of far-reaching and rather convoluted impacts begins with actions in the U.S. to protect the northern spotted owl in the early 1990s. The prescribed solution required sharp reductions in timber harvesting in Washington, Oregon, and northern California, with little consideration of where replacement timber might come from or what the impact might be on wildlife species in the new producing regions. As harvests in the U.S. declined, the flow of logs and timbers from the U.S. Pacific Northwest to Japan and other points in the Far East slowed. Conversely, the flow of softwood lumber from Canada to the U.S. increased. As Canadian shipments of wood to the U.S. grew, the Canadians were able to ship less to Japan, Korea, and Taiwan. These countries, in turn, began to seek a more reliable trading partner and new sources of softwood supply, looking to Russia, and in particular the Russian Far East for softwood logs and lumber. The forests of the Russian Far East, it turned out, were far less productive than those of the Pacific Northwest, requiring the harvest of 1.6–1.9 times greater forest area to produce the same volume of wood per annum as in the Pacific Northwest, and under far less stringent harvesting regulations. Shortly thereafter—in early fall 1995—the cover of *Time* magazine carried the title "The Rape of Siberia" and articles within expressed concerns about growing interest on the part of Japan, Korea, and the U.S. in the rich forests there and about the effects of rising forest harvest activity on the long-term environmental health of what was described as a pristine and fragile region.

Whether banning industrial activity for purposes of protecting endangered wildlife without any consideration of whether there are rare or endangered species in potential new producing regions, or designating lands as preserves with the intent of blocking imminent or future raw material procurement with no thought of where needed resources might come from, or systematically erecting regulatory and economic barriers to domestic resource extraction, the effect is the same: to shift undesirable impacts of domestic consumption to somewhere else. A Harvard University research team summarized the situation this way:

> *The United States and other affluent countries consume vast quantities of global natural resources, but contribute proportionately less to the extraction of many raw materials. This imbalance is due, in part, to domestic attitudes and policies intended to protect the environment. Ironically, developed nations are often better equipped to extract resources in an environmentally prudent manner than the major suppliers. Thus, although citizens of affluent countries may imagine that preservationist domestic policies are conserving resources and protecting nature, heavy consumption rates necessitate resource extraction elsewhere and oftentimes under weak environmental oversight. A major consequence of this "illusion of natural resource preservation" is greater global environmental degradation than would arise if consumption was reduced and a larger portion of production was shared by affluent countries. Clearly, environmental policy needs to consider the global distribution and consequences of natural resource extraction.*

Considering Ethical Standards

Questionable standards exist not only in resource procurement, but also in regard to disposal of unwanted wastes. Thinking, however, appears to be similar in both cases. Consider the position espoused by Lawrence Summers, then the chief economist of the World Bank, in a 1992 internal memo. Summers argued that international trade in toxic wastes was a good thing since it provided a mechanism for removal of unwanted wastes from countries with high environmental standards and provided countries with low wages and low population densities with jobs and income. He also suggested that workers in low income countries pose lower economic risks of wage loss due to illness [because wages are quite low to begin with], and thus possess a higher tolerance for environmental hazard. When the memo was leaked and printed, in part, in *The Economist,* a firestorm of negative public opinion erupted around the world, from developed and developing countries alike. Economists and others later defended the logic articulated by Summers as basic economics. As one writer explained "Because of their higher incomes, people in developed countries can afford, and are willing to pay, a higher price to dispose of these pollutants in a way that reduces risk to themselves. Also, being affluent, they can easily forgo further increases in income that industries generating these wastes might offer. The economic way of stating this relationship is that citizens of the developed world have a high opportunity cost in increases of pollution."

From an ethical perspective a large segment of global society found Summers' position to be unacceptable. So if export of hazardous wastes, that might provide jobs and income but also create significant environmental and health risks, is clearly unacceptable, then what about mass-scale importation of raw materials that result in jobs and income to other countries but also create substantial long or short-term environmental degradation and perhaps significant resource depletion within those countries? These kinds of arrangements appear to be far less controversial than exporting garbage or hazardous wastes as evidenced by the fact that they are made on a daily basis with virtually no reaction or even notice on the part of citizens or governments of wealthy importing countries.

Focusing on Ecosystems—But Not on Consumption

Beginning in the early 1980s, and given a substantial boost by the Clinton/ Gore administration, the management objective of public lands within the U.S. changed so as to focus on such things as biodiversity, recreation, and aesthetics and to sharply reduce commodity production. It was a dramatic shift.

What follows, with the author's permission, is the entire text of a presentation delivered by Douglas MacCleery at the "Building on Leopold's Legacy" conference in Madison, Wisconsin, on October 4, 1999. In the text MacCleery, then Assistant Director of Forest Management of the USDA-Forest Service, refers several times to Aldo Leopold, who is considered by many to be the father of wildlife management and the U.S. wilderness system. As explained by the Foundation that now bears his name, Leopold was "a conservationist, forester, philosopher, educator, writer, and outdoor enthusiast." MacCleery's presentation provides another example of the startling disconnect between consumption and the environment in American society.

In the words of MacCleery (through top of page 32):

> *Over the last two decades there has been a substantial shift in the management emphasis of public lands, particularly federal lands, in the United States. That shift has been to a substantially increased emphasis on managing these lands for biodiversity protection and amenity values, with a corresponding reduction in commodity outputs. Over the last decade, timber harvest on National Forest lands has dropped by 70 percent, oil and gas leasing by about 40 percent, and livestock grazing by at least 10 percent.*
>
> *Terms like "ecosystem management," an "ecological approach to management," and, more recently, "ecological sustainability" have been used to describe this change in the management emphasis of public lands. Many have referred to it as a significant "paradigm shift." Just recently, a Committee of Scientists issued a report proposing that the National Forests be managed for "ecological sustainability," where primary*

management emphasis is to be placed on "what is left" out on the land, rather than "what is removed." Commodity outputs, if they are produced, would become a derivative or consequence of managing National forests for primarily a biodiversity protection objective. Significantly, some Committee members justified this recommendation in part on "ethical and moral" grounds.

Many have attributed the move to ecosystem management or ecological sustainability to a belated recognition and adoption of Aldo Leopold's "land ethic"—the idea that management of land has, or should have, an ethical content. This year, celebrations are planned commemorating the 50th anniversary of the publishing of Leopold's A Sand County Almanac, *in which he spoke eloquently about the need for an ethical obligation toward land use and management. One sign that Leopold's ideas have finally struck a chord with the larger society is that conservation issues are increasingly being taken up as causes of American churches.*

While a mission shift on U.S. public lands is occurring in response to changing public preferences, that same public is making no corresponding shift in its commodity consumption habits. The "dirty little secret" about the shift to ecological sustainability on U.S. public lands is that, in the face of stable or increasing per capita consumption in the U.S., the effect has been to shift the burden and impacts of that consumption to ecosystems somewhere else. For example, to private lands in the U.S. or to lands of other countries.

Between 1987 and 1997, federal timber harvest dropped 70 percent, from about 13 to 4 billion board feet annually, which translates into about a one-third reduction in U.S. annual softwood lumber production. A significant effect of this reduction, in the face of continuing high levels of per capita wood consumption, has been to transfer harvest to private forest ecosystems in the U.S. and to forest ecosystems in Canada. For example:

- *Since 1990, U.S. softwood lumber imports from Canada rose from 12 to 18 billion board feet, increasing from 27 to 36 percent of U.S. softwood lumber consumption. Much of the increase in Canadian lumber imports has come from the native old-growth boreal forests. In Quebec alone, the export of lumber to the U.S. has tripled since 1990. The increased harvesting of the boreal forests in Quebec has become a public issue there.*
- *Harvesting on private lands in the southern United States also increased after the reduction of federal timber in the West. Today, the harvest of softwood timber in the southeastern U.S. exceeds the rate of growth for the first time in at least 50 years. Increased harvesting of fiber by chip mills in the southeastern U.S. has become a public issue regionally.*

Today the U.S. public consumes more resources than at any time in its history, and it also consumes more per capita than almost any other nation. Since the first Earth Day in 1970, the average family size in the United States has dropped by 16 percent, while the size of the average single family house being built has increased by 48 percent.

The U.S. conservation community and the media have given scant attention to the "ecological transfer effects" of the mission shift on U.S. public lands. Any ethical or moral foundation for ecological sustainability is weak indeed unless there is a corresponding focus on the consumption side of the natural resource equation. Without such a connection, ecological sustainability on public lands is subject to challenge as just a sophisticated form of NIMBYism ("not in my backyard"), rather than a true paradigm shift.

A cynic might assert that one of the reasons for the belated adoption of Aldo Leopold's land ethic is that it has become relatively easy and painless for most of us to do so. When Leopold was a young man forming his ideas, more than 40 percent of the U.S. population lived on farms. An additional 20 percent lived in rural areas and were closely associated with the management of land. Today less than two percent of us are farmers and most of us, even those living in rural areas, are disconnected from any direct role in the management of land. Adopting a land ethic is easy for most of us today, because it imposes the primary burden "to act" on someone else.

While few of us are resource producers any more, we all re-main resource consumers. This is one area we all can act upon that could have a positive effect on resource use, demand and management. Yet few of us connect our resource consumption to what must be done to the land to make it possible. At the same time many of us espouse the land ethic, our operating motto in the marketplace seems to be "shop 'till you drop" or "whoever dies with the most toys wins."

The disjunct between people as consumers and the land is reflected in rising discord and alienation between produc-ers and consumers. Loggers, ranchers, fishermen, miners, and other resource producers have all at times felt themselves subject to scorn and ridicule by the very society that benefits from the products they produce. What is absent from much environmental discourse in the U.S. today is a recognition that urbanized society is no less dependent upon the products of forest and field than were the subsistence farmers of America's past. This is clearly reflected in the language used in such discourse.

Rural communities traditionally engaged in producing timber and other natural resources for urban consumers are commonly referred to as natural resource "dependent" communities. Seldom are the truly resource dependent communities like Boulder, Denver, Detroit, or Boston ever referred to as such.

One of the relatively little known aspects of Aldo Leopold's career is the years he spent at the Forest Service's Forest Products Lab at Madison, Wisconsin. While there, he spoke of the need for responsible consumption. In 1928 Leopold wrote:

"The American public for many years has been abusing the wasteful lumberman. A public which lives in wooden houses should be careful about throwing stones at lumbermen, even wasteful ones, until it has learned how its own arbitrary demands as to kinds and qualities of lumber, help cause the waste which it decries. . . .

The long and the short of the matter is that forest conservation depends in part on intelligent consumption, as well as intelligent production of lumber."

If management of land has an ethical content, why does not consumption have a corresponding one, as well? Is there a need for a "personal consumption ethic" to go along with Leopold's land ethic? In his wonderful land ethic chapter in A Sand County Almanac, Leopold wrote that evidence that no land ethic existed at the time was that a "farmer who clears his woods off a 75 percent slope, turns his cows into the clearing, and dumps its rainfall, rocks, and soil into the community creek, is still (if otherwise decent) a respected member of society."

To take off on that theme and make it more contemporary, the evidence that no personal consumption ethic exists today is that a "suburban dweller with a small family who lives in a 4000 square-foot home, owns three or four cars, commutes to work alone in a gas guzzling sport utility vehicle (even though public transportation is available), and otherwise leads a highly resource consumptive lifestyle is still (if otherwise decent) a respected member of society. Indeed, her/his social status in the community may even be enhanced by virtue of that consumption."

Ecosystem management or ecological sustainability on public lands will have weak or non-existent ethical credentials and certainly will never be a truly holistic approach to resource management until the consumption side of the equation becomes an integral part of the solution, rather than an afterthought, as it is today. Belated adoption of Leopold's land ethic was relatively easy. The true test as to whether a paradigm shift has really oc-curred in the U.S. will be whether society begins to see personal consumption choices as having an ethical and environmental content as well—and then acts upon them as such.

Ecosystem management remains the dominant philosophy in public land management, but as in 1999, there is as yet no movement toward changing the philosophy of consumption. Meanwhile, the nation's largest environmental organization—the Sierra Club—is doing everything it can to reduce harvesting activity to zero in the 192 million acres of federal National Forests, in effect seeking to change them into national parks. A significant segment of the public appears to support this position.

Justifying Transfer of Environmental Impacts

Two beliefs, one environmental in nature and one economic, justify the transfer of environmental impacts to developing countries. First is the be-lief that environmental protection in the U.S. and other highly developed countries is of paramount importance, a priority that often leads to initia-tives to prevent or minimize industrial activity. Second is the belief that it makes economic sense for all concerned for developed countries to rely on developing countries as sources of basic raw materials, sites of labor-intensive manufacturing goods, and even repositories of hazardous wastes. This view is consistent with the economic principle of comparative ad-vantage and is basic to the demographic transition phenomenon discussed in Chapter 12. Both views support the transfer of heavy industrial activity and associated environmental impacts from the most economically devel-oped to less economically developed nations.

The economic justification for transferring high impact activity to de-veloping countries is that the citizens of these countries come out ahead even when environmental impacts are high, because jobs and income ac-company those impacts. The developed countries also benefit since com-panies within developing countries are often able to produce products at lower cost than they can be produced elsewhere, commonly because of lower wages and regulatory restrictions on such things as emissions to the environment—in economic terms, such countries are said to have comparative advantage in producing certain kinds of products. The result is something of an economic "win-win" situation.

With respect to the principle of comparative advantage, consider for a moment the 1986 saga of the *Khian Sea,* a vessel that sailed from Philadelphia with a load of 13,000 tons of toxic incinerator ash bound for Haiti. After the government there denied permission to off-load, the vessel sailed (minus 3,000 tons of ash that was left on the Haitian beach) for 27 months in search of a country that would accept its cargo. Ultimately, the ship docked in Shanghai absent its unwanted cargo. It was never determined where the waste wound up. This incident focused attention on the common practice of hazardous waste export from developed to developing countries, a practice that continues today. Several years after the *Khian Sea* incident, Lawrence Summers penned the internal World Bank memo referred to earlier which said that international trade in toxic wastes was a good thing for developed and developing countries alike. Two views: to economists—basic economics; to many within the developing nations and elsewhere—an ethically bankrupt position.

As a postscript, in 2010, at the 10th Conference of the Basel Convention, 178 countries voted to support a ban on toxic waste exports (and particularly electronic or "E" wastes) to developing countries. The U.S. was one of 17 countries not supporting the measure (and at this writing still does not), although the U.S. does vigorously prosecute the *illegal* export of such waste. Bills to ban overseas dumping of E-wastes were introduced in the U.S. House and Senate in 2013 and 2014, and as of late 2015 25 states had enacted laws mandating state-wide E-waste recycling. Meanwhile, the U.S. remains the world's leading exporter of electronic waste.

References

Alden, A. 2013. "California's Looming Sand and Gravel Crunch." *KQED Science,* June 6.

Anonymous. 1992. "Let Them Eat Pollution." *The Economist,* 7745: 82, Feb. 8.

B Corp. 2016. "Using Business as a Force for Good," Accessed January 6,
      2016. (http://www.bcorporation.net/)

Berlik, M., D. Kittridge, and D. Foster. 2002. "The Illusion of Preservation:
      a Global Environmental Argument for the Local Production of Resources."
      Journal of Biogeography 29, 1557–1568.

Bradley, L. 2014. "E-Waste in Developing Countries Endangers Environment,
      Locals." *U.S. News and World Report,* August 1.

Clinkenbeard, J. 2012. *Aggregate Sustainability in California.* California
      Geological Survey, Department of Conservation.

Coalition for Environmentally Responsible Economies (CERES). 2006.
      "Investors Press S&P 500 Companies for Better Social, Environmental
      Disclosure." *PR Newswire,* Press release, Oct. 5.

Evans, A., R. Everett, S. Stephens, and J. Youtz. 2011. *Comprehensive Fuels
      Treatment Practices Guide for Mixed Conifer Forests: California, Central
      and Southern Rockies, and the Southwest.* Forest Guild/U.S. Forest Service.

Gorman, A. 2003. "Mexican Officials Dig In Their Heels Over Mining of Baja
      Sand." *Los Angeles Times,* March 25.

Johnson, J., G. Pecquet, and L. Taylor. 2007. "Potential Gains from Trade in
      Dirty Industries: Revisiting Lawrence Summers' Memo." *Cato Journal*
      27(3): 397–410.

Kinsley, M. 2008. "Revisiting One Lawrence Summers Controversy."
      Washington Post, Nov. 8.

Knudson, T. 2003a. "California—the State of Denial." *Sacramento Bee,* April 27.

_____ . 2003b. "Grounds for Anger. State of Denial." *Sacramento Bee,* August 17.

Laaksonen-Craig, S., G. Goldman, and W. McKillop. 2003. *Forestry, Forest
      Industry, and Forest Products Consumption in California.* University of
      California Division of Agriculture and Natural Resources, Publication
      8070.

Lewis, D., and R. Chepesiuk. 1994. *The International Trade in Toxic Waste: A Selected Bibliography of Sources.* Rock Hill, South Carolina: Winthrop University.

Linden/Yakutsk, E. 1995. "Siberia: The Tortured Land." *Time,* Sept. 4.

MacCleery, D. 1999. "Aldo Leopold's Land Ethic: Is it Only Half a Loaf Unless a Consumption Ethic Accompanies It?" *Forest Management Update,* 20: 32–34.

McAllister, L. 2013. "The Human and Environmental Effects of E-Waste." *Population Reference Bureau,* April.

McKillop, W. 1995. *Industrial Forestry and Environmental Quality.* University of California, Berkeley, S.J. Hall Lecture.

Naisbitt, J. 1988. *Megatrends: Ten New Directions Transforming Our Lives.* New York: Warner Books.

Perez-Garcia, J. 1993. *Global Forestry Impacts of Reducing Softwood Supplies from North America.* University of Washington, Center for International Trade in Forest Products, Working Paper #43.

Stewart, W., R. Powers, K. McGown, L. Chiono, and T. Chuang. 2011. *Potential Positive and Negative Environmental Impacts of Increased Woody Biomass Use for California.* California Energy Commission. Publication Number: CEC-500-2011-036.

Thompson, R., and C. Dicus. 2005. *The Impact of California's Changing Environmental Regulations on Timber Harvest Planning Costs.* California Institute for the Study of Specialty Crops/ Forest Foundation, March.

U.S. Department of the Interior. 2002. "Department of the Interior and Related Agencies Appropriations Act, 2003." *Department of the Interior and Related Agencies Appropriations Act, 2003, Statement of U.S. Representative Lois Capps.*

Vandergert, P., and J. Newell. 2003. "Illegal Logging in the Russian Far East and Siberia." *International Forestry Review* 5(3): 303–306.

CHAPTER 3

Misconceptions

It ain't what you don't know that gets you into trouble,
it's what you know that just ain't so.

—MARK TWAIN

Misconceptions can lead to nonsensical and irresponsible behavior. Mistaken beliefs, misunderstandings, and views or opinions that are incorrect because they are based on faulty thinking can and do lead to harm and human suffering. Misconceptions benefit no one except occasionally those who may have an agenda. Examples are plentiful.

A Personal Experience

Teaching in a university setting provides an avenue for gaining insight into the knowledge, thought processes, and perceptions of a broad cross-section of society. College students today are remarkably frank, and classroom dynamics are such that very little is accepted as fact without questioning, comment, and discussion. This is particularly true when courses are specifically organized to encourage and facilitate discussion. And so it was in the winter of 1991 when, at the University of Minnesota, I began a discussion of forests, forestry, and forest harvesting in a new class entitled "Natural Resources as Raw Materials." What happened, in short, was that questions from a large segment of the class quickly became not only extraordinarily challenging, but in some cases outright hostile—a situation I had not encountered in over 20 years of teaching.

A bit shaken by the end of the class session, my mind was racing as I gathered up my things and walked slowly through the crisp, cold air toward my office. Where were these students coming from? Why did the topic of forestry and forest harvesting trigger such strong emotions? Were the gross misconceptions about forests that had been expressed by several students limited to only those who had spoken up, or to others as well? Determined to find out, I sat down that same afternoon and devised a quiz for the next class meeting that was designed to assess knowledge about natural resources, forests and other environmentally-related matters. The responses to that quiz revealed a pervasive pessimism. The students consistently and overwhelmingly selected answers indicating the environmental situation to be worse than it really was, and a shocking level of misinformation regarding natural resources, forests, and of domestic forests in particular.

As luck would have it, only two weeks after that eye-opening class-session a rambling article "The Fate of the Forest" appeared in *City Pages,* an entertainment-oriented downtown newspaper of Minneapolis. The article was devoid of any kind of accurate factual information. In addition it indicated an appalling level of ignorance and cynicism about economics, and specifically about the role of profit in American business and industry, and how basic industries such as mining and logging contribute to our everyday lives. The article concluded with a quote from an ultra-left-wing radical who described loggers as "a rot, a disease, and an aberration against nature . . ." So, I dutifully took this article into my class to see what they might think about it. What emerged after a bit of questioning and nudging was that while virtually nobody agreed with the tone of the article, a significant number had a negative view of the term "profit." When I inquired, for instance, what reasons an individual might have for going into business as a manufacturer of cement or of lumber, many students' responses included the word "greed," ignoring altogether the reality of rising consumption of cement and lumber within the U.S.

Several months later I attended an international conference focused on wood and materials science. In talking with colleagues from universities across North America, I shared the quiz results from the University of Minnesota. By the end of that conference a number of professors indicated an interest in testing their students. So, in 1992 the environmental quiz was administered to about 2,000 students at eleven major universities across North America. With the exception of one university (where it appeared that questions may have been discussed with students prior to the quiz), results were remarkably similar in all regions, and almost identical to those obtained in the original test of University of Minnesota students. Student perceptions of environmental conditions were consistently pessimistic and at wide variance with reality.

As an example of responses from the students surveyed:

- 50 percent believed that the United States is a net exporter of most raw materials used by industry today. [*The United States was then and remains a net importer of every category of industrial raw material—metals, cements, petroleum (the basis for most plastics), and wood—and in many cases by a substantial margin*].

- 78 percent agreed with the statement "The world is rapidly running out of many important minerals." [*This is the one statement that has an element of truth to it. It is true that the mid- to long-term availability of several important minerals is threatened by increasing consumption, environmental constraints, and modest investment globally in exploration, metals processing, and recovery and recycling. However, the world is generally not running out of minerals. Because the Earth's crust is composed of a vast array of minerals, the world will likely never "run out" of them.*].

- 94 percent underestimated the percent of annual U.S. paper production that is produced from recycled paper. Some 71 percent, in fact, estimated U.S. recycled paper production to be less than one-half of what it really is, and 45 percent of respondents estimated it to be one-fifth or less of actual. [*Recovery of wastepaper for domestic recycling and export totaled more than 38.6 million tons in 1994, amounting to 40.3 percent of domestic production. Of the paper recovered, 30.0 million tons were recycled in U.S. paper mills and 8.6 million tons were exported. This translated to a U.S. wastepaper utilization rate of 32.9 percent in 1994. As of 2015 the U.S. wastepaper recovery rate was 66.8* percent].

- 65 percent indicated that forest harvest exceeds net growth in U.S. forests. [*The opposite has been true for every year since at least the early 1950s, and likely for two decades before that*].
- 73 percent agreed with the statement "At current rates of deforestation, 40 percent of current forests in the U.S. will be lost by the middle of the next century." [*Then, as now, the area covered by forests in the United States was increasing. While some forests were being lost to urban expansion, highway expansion, and other infrastructure development, lands cleared for agriculture a century or two earlier and subsequently abandoned were steadily returning to forest cover, and at a faster rate than forest loss. As a result, forest cover has remained stable for more than a century and today is slightly greater than it was in the early 1900s*].
- 76 percent underestimated the percentage of area currently covered by forests in the U.S., compared to forest coverage in 1600. [*U.S. forests today cover an area equal to about 72 percent of that covered by pre-settlement forests (3), but 76 percent of the respondents chose a percentage that was one-half or less of the actual percentage*].

These results, that were almost identical to those of my class, clearly show why many of the students were outwardly hostile to anyone talking about the harvesting of forests or extraction of minerals. A large segment of them believed that the United States was in the process of being deforested, that little or no paper was being recycled, and that minerals were at the brink of extinction.

Since 1992 the "environmental quiz" has been given to thousands of high school and middle school students, boy scouts, and various groups of adults, including industry, government, and community leaders. And, I continued to quiz students at the University of Minnesota every year through 2006. Although older respondents tended to score better than youngsters and young adults, responses nonetheless typically varied significantly from reality and reflected a pessimistic viewpoint. Responses from young adults up through 2006 remained nearly identical to the responses of the college students in 1991 and 1992.

After the winter of 1991, I fundamentally changed my approach to teaching of environmental issues. I began to take less for granted, and to focus part of my teaching on what is and isn't true. I also made room for discussion of things such as the global economy and global competitiveness issues, consumption trends, trade imbalances and why they matter, and the U.S. balance of payments deficit—topics that typically would receive no attention in a class dealing with the environment.

Several people with whom I have discussed the matter of student misconceptions have suggested that such a view of the world isn't necessarily bad, and that a heightened if inaccurate view of problems might even be good as it could spur them to action. I could not disagree more. The young people who populate high school and college classes will become adults and societal leaders within a very short time frame. I am reminded that those students of 1991 are now in their low to mid-50s, and that all are voters, with most holding positions in which they influence others (including their own children). Further, the coming years will see them move increasingly into positions of authority within society. I am also reminded that in politics as in everyday life, it is not what is true that guides decision making. It is what is *perceived* to be true that carries the day.

Misconceptions matter. In the face of strongly held misconceptions, problem solving becomes difficult to impossible. In fact, even defining problems becomes difficult. Gaining acceptance of new knowledge is challenging. Rational decision-making is stymied. For example, with regard to raw material needs and the environment, people who believe that the world is on the brink of running out of raw materials are likely to take radically unrealistic positions. Should the U.S. and the world take steps to drastically inhibit mining activity? This is not such a radical question if it is believed that next week is likely to mark the last shovelful of ore. Similarly, someone who believes that each tree harvested is a step toward deforestation of the United States is also far less likely to be part of a rational discussion about forest management than an individual who understands forest dynamics and trends. Approaches to environmental protection, or any other endeavor for that matter, which are based on misconceptions are unlikely to have beneficial outcomes.

Misconceptions and Tragedy

In the late 1500s, Italian Philosopher Giordano Bruno proposed that the stars in the heavens were in reality similar to the sun, and that other planets revolved around them. He also subscribed to the view, advanced several decades earlier by Copernicus, that the Earth revolved around the sun rather than the Earth being at the center of the universe. While not an astronomer, Bruno nonetheless helped to establish a foundation for thinking independent of authoritarianism, and his speculations have been described as "a lasting and significant contribution to our modern conceptualization of a dynamic universe." However, in part because of these views, which conflicted at the time with accepted truth, Bruno was tried and convicted of heresy and other charges and subsequently burned at the stake.

Just three decades thereafter, Galileo, who is credited with major accomplishments in astronomy, confirmed in his writings the work of Copernicus relative to the place of Earth in the universe, acknowledging that the sun occupies the center of the universe and that the Earth revolves around it. In 1633 he too was convicted of heresy, and he spent the last nine years of this life under house arrest. The effect was to further delay the pursuit of knowledge and study of astronomy.

Misconception also fueled the witch hunts of the mid-1500s to late 1600s that spread across Europe and then America. Based on beliefs that some in society were possessed by demons, people were denounced, arrested, and in some cases executed. Merely being left-handed was in some instances deemed sufficient evidence to identify a person (most often a woman) as a witch.

Left handedness was also for more than two centuries, even into the early 20th century, used as a reason for discrimination and sometimes institutionalization in Western Europe and North America. Soviet bloc countries continued such practices into the 1970s, and Albania actually made left-handedness illegal, and punishable as a crime. Because it was "known" that writing or dominantly using the left hand was abnormal and therefore wrong, educators tied children's left hands behind them and inflicted deliberate pain to force right-handedness. Others pointed to left-handedness as an indication of perversity, savagery, and criminality. Many suffered and society was diminished as a result.

In the mid-1800s, Hungarian physician Ignaz Semmelweis noted that the death rate in childbirth was three times higher in doctor-assisted childbirth than when midwives working in the same hospital attended to mothers as they gave birth. Based on his observations while working as an assistant in the maternity clinic in Vienna General Hospital, Semmelweis encouraged interns who had recently performed autopsies to wash their hands with chlorinated lime solutions before assisting in childbirth. The effect was immediate and dramatic, reducing the incidence of fatal puerperal fever from about 10 percent to about 1–2 percent. Unfortunately, some doctors were offended by the notion that hand washing could reduce mortality rates as this conflicted with what was "known" at the time. As a result, Semmelweis' findings were largely ignored by the medical community in Vienna. Prominent obstetricians and physicians abroad also mostly rejected his recommendations. Fifty years later, sterile surgery, covering of wounds, and hand washing were introduced by Dr. Joseph Lister, based on Louis Pasteur's research in the field of microbiology, and such practices were quickly adopted by the medical profession.

Dr. Ignaz Semmelweis. 1818–1865

Rigid adherence to long-held misconceptions in this case resulted in countless unnecessary deaths of mothers and newborns and misguided persecution of a pioneer in medicine. Years later, this experience led to definition of the "Semmelweis reflex" or "Semmelweis effect" to describe the reflex-like tendency to reject new evidence or new knowledge because it contradicts established norms, beliefs or paradigms.

Today, misconceptions are causing people to reject vaccination of their children, with the result that diseases long ago conquered are re-emerging, endangering not only the specific children involved, but all of society. In west and central Africa, myth and superstitions are hampering efforts to halt the spread of the Ebola virus.

In some ways, current resistance to growing evidence of climate change mirrors the Semmelweis experience, but with an important difference. Semmelweis was not a scientist, and did not have a theoretical or scientific basis for his claims. What he advocated was not accepted until scientific findings made it impossible to ignore the role of germs in the spread of disease. However, in the case of climate change, verifiable measurements by numerous institutions over a long period of time, coupled with rigorous studies conducted by thousands of scientists, leave little doubt that the Earth's climate is warming and that fossil carbon is playing a major role. As with hand washing, there is enormous resistance to suggestions that long established climate norms may be changing, and that long established human conduct may need to change as well. But this time, the existence of an extensive scientific underpinning has not ended debate or resistance to change. As time passes, the potential for harm and substantial human suffering is rising, and it will be left to future generations to record the conclusion of this chapter of human history.

Misinformation and the Environment

Misconceptions are not limited to astronomy and astrophysics, human traits, medicine, and climate dynamics. Mistaken beliefs regarding almost all aspects of the environment and environmental trends, including climate change, are numerous and pervasive, stemming in large part from deliberate spreading of misinformation. Examples abound of misinformation generated by industry for purposes of downplaying product risks or influencing public policy. Numerous examples can also be found of deliberate promulgation of misinformation, exaggeration, and deception on the part of environmental advocates. Misconceptions may also arise from well-intended, but ill-informed, instruction and from efforts that unintentionally trivialize complex environmental problems.

Industry and Misinformation

In selling, there is an almost irresistible temptation to tout and even exaggerate a product's favorable attributes, while downplaying or omitting mention of less favorable characteristics. That is why the phrase *caveat emptor* (from Latin—literally "let the buyer beware") was coined in the early 16th century. The majority of consumers are well aware of the tendency to overstate, and consequently tend to take what a salesperson tells them with a grain of salt. Many take time to investigate purchases ahead of time to determine factual information. Thus, while misinformation is common in the selling process, harm is minimized because consumers know to be cautious.

A more modern, and darker, phenomenon is coordinated dissemination of misinformation and half-truths for purposes of influencing public policy. The tobacco industry is a poster child for this approach. David Michaels, in his 2008 book *Doubt is Their Product,* outlined how the tobacco industry worked to skew scientific literature. That industry raised doubts about indisputable scientific evidence, by paying credentialed spokespeople to dispute findings of other scientists and research organizations. They coined the term "junk science" to dismiss peer reviewed research and "sound science" to elevate studies created by the product defenders. In addition, they funded lobbying focused on disputing scientific evidence about health risks associated with smoking. This initiative was sufficiently successful to delay by decades any meaningful action to address tobacco promotion and use.

As reported by the Union of Concerned Scientists, a similar effort was begun in 1998 by Exxon-Mobil, ostensibly patterned after the strategy of the tobacco industry, to create doubt regarding climate change. Key elements of this effort were threefold:

- raising doubts about virtually every piece of established scientific evidence,
- using seemingly independent front organizations to disseminate messages to lend credibility and to confuse the public, and
- publishing and republishing non-peer reviewed research reports authored by a small cadre of scientific spokespeople containing evidence that conflicted with research published in mainstream scientific journals.

One organization, that continues to operate as of this writing, specializes in disseminating "facts" about climate change. One tactic employed in recent years has been on-line publication of news articles with embedded links to highly regarded research organizations. Often, however, what has been reported by the research organizations referenced is at wide variance with the "news" that is reported. This strategy is apparently based on an assumption that few will go to the sources cited to see what is actually reported. Overall, this effort also appears to have been (and continues to be) highly successful—so much so in fact, that hundreds of millions in the United States alone today doubt the findings of climate scientists. A subset of this group appears to have completely rejected science and the scientific method in general.

While industry or any other organization has every right to publish credible material that raises public awareness and fosters discourse about a particular topic, a line is crossed when efforts shift to misinformation campaigns. That line is doubly crossed when the misinformation deliberately attempts to destroy public trust in science. As in every instance of misinformation, no one benefits except perhaps the entity seeking to muddy the waters. And to the extent that trust in the scientific method is lost, all of society loses.

Environmental NGOs and Misinformation

Seeking to subvert science when scientific findings fail to support a specific agenda is clearly not limited to the private sector. Several prominent environmental non-governmental organizations (ENGOs) have long been playing this game as well.

One of the prominent voices decrying the misconduct of some environmental ENGOs is Dr. Patrick Moore, a co-founder of and 15-year activist with Greenpeace. Moore charges that Greenpeace, and the environmental movement in general, has lost its way, resorting to misinformation and scare tactics to further its agenda and fundraising efforts on a wide range of issues ranging from genetic enhancement and chlorine in drinking water to wind power and forestry. Similar charges have been leveled by Gregg Easterbrook, Michael Fumento, Bjørn Lomborg, and others. Fumento spotlighted what he called an assault on science in environmentalist campaigns targeting the pesticide Alar, food irradiation, and other perceived threats. Like industry, ENGOs lobby heavily, and in recent years such groups have increasingly produced poorly-researched, often inaccurate reports on various topics, and then cited these in other such reports, creating a basis for arguing that there are two sides to issues for which science clearly points to different conclusions.

Perhaps taking a cue from climate change disinformation campaigns, another tactic that is increasingly employed is release of bogus information accompanied by claims that the origin is a credible source. A recent example comes from the Bureau of Ocean Energy Management (BOEM) which issued a public disavowal of statements made by a group of environmental activists who were seeking to drum up support for discontinuation of use of seismic air guns in underwater geological exploration. Six months earlier BOEM had issued a statement saying "To date, there has been no documented scientific evidence of noise from air guns used in geological and geophysical (G&G) seismic activities adversely affecting animal populations." In response, a web posting soon appeared saying "Seismic air gun testing currently being proposed in the Atlantic will injure 138,000 whales and dolphins and disturb millions more, according to government estimates." In its disavowal, BOEM strongly indicated that this and similar statements are not accurate, and that they misrepresent estimates of agency scientists. However, as of this writing, the statement in question continues to be promulgated.

The human tendency to exaggerate is accentuated not only by sellers, but also by those seeking donations. The story can always be made a little more compelling, a bit more urgent by adding a bit of inflammatory language here, omitting critical facts there, and doing whatever possible to create a sense of crisis or urgency. Careful examination of fundraising letters, and investigation of the problem being discussed, shows many of them to be less than truthful in one way or another. This practice apparently works well, as it is repeated again and again. One fundraising letter from a large U.S. based environmental organization began with the words "Bulldozers Are Advancing on the National Forests," this at a time when timber harvests within the national forests were near their lowest level in history. The specter of a rumbling formation of large yellow machines descending on the forests must have been quite effective, since the same letter was sent to prospective donors multiple times over a period of several years.

Education and Environmental Misinformation

The very institutions entrusted with environmental education may also be spreading misinformation. Hopefully, such misinformation is unintentional. However, damage is done regardless of intention.

A mid-1990s study of science textbooks used in U.S. schools and of environmental books for children found numerous errors of fact and exaggeration of environmental problems and trends. Another review of K-12 teaching materials at about the same time concluded that "factual errors are common in many environmental education materials and textbooks." This report also noted that "many high school environmental science textbooks have serious flaws . . . [including a propensity to] mix science with advocacy." It was further noted that educational materials often fail to prepare students to deal with controversial issues or to help students understand tradeoffs in addressing environmental problems. While there have been no recent studies of the accuracy of environmental information in student textbooks, spot checks of such texts suggest ongoing problems.

Misinformation or creation of misconceptions can come about in more ways than via false or misleading information. A 2000 investigation that included focus group sessions with teachers centered on understanding one specific issue—the sources of substantial misinformation about tropical deforestation. Focus group sessions revealed that some of this information was being conveyed in classrooms, sometimes through reliance on activist literature as sources of information, with other messages likely conveyed in the course of computer-based self-study. An interesting finding was that many teachers, because students tend to find the deforestation topic discouraging, try to do something at the conclusion of discussion to allow students to "feel good." Examples of such activities include giving pennies to buy up forested acreage somewhere in the world, or to send a letter to someone somewhere about this problem. Investigators noted that such activities mislead students by overlooking completely the complexity of factors underlying deforestation. These include population growth, conversion to permanent agriculture, shifting agriculture, poverty, lack of land ownership, logging, mining, and more. This poses the question: to whom would one write about taking action on these matters?

A similar problem can arise from well-indended initiatives such as selling T-shirts or attending a rock concert to benefit the tropical forests. While such activity may allow young people to feel good, it also trivializes the problem and sends a message that a few minutes of inspired activism can materially help to solve complex environmental problems. Instead, perhaps, the message should be that the world needs people educated in science, ecology, agriculture, forestry, social sciences, and related disciplines who have good foreign language skills, and who are interested in working with others to find solutions.

Science and Misconceptions

That science is important to society is undeniable. A succinct summary of the role of science in society was included in a July 1945 report to President Harry Truman, entitled, "Science: The Endless Frontier." In it, author Dr. Vannevar Bush, Director of the Office of Scientific Research and Development, said in part:

> *Advances in science when put to practical use mean more jobs, higher wages, shorter hours, more abundant crops, more leisure*

for recreation, for study, for learning how to live without the deadening drudgery which has been the burden of the common man for ages past. Advances in science will also bring higher standards of living, will lead to the prevention or cure of diseases, will promote conservation of our limited national resources, and will assure means of defense against aggression. But to achieve these objectives—to secure a high level of employment, to maintain a position of world leadership—the flow of new scientific knowledge must be both continuous and substantial . . . Science, by itself, provides no panacea for individual, social, and economic ills . . . But without scientific progress no amount of achievement in other directions can insure our health, prosperity, and security as a nation in the modern world.

Scientific inquiry is employed to shed light on things unknown, to provide a foundation for deeper understanding, and to overcome misconceptions that may be inhibiting consideration or acceptance of new ways of doing things. In the absence of science, as was the case 400 years or so ago, progress and new knowledge is reliant on real time trial and error, with solutions to and understanding of difficult, complex problems and phenomena out of reach of society. So when some today, even some who aspire to positions of leadership, proclaim that they don't believe in science, it is fair to consider the alternative. Inevitably, purveyors of rumor, unsubstantiated conjecture, mysticism, and dogma would rise in stature, and society would be the poorer. To understand the consequences of a world without scientific inquiry, we have only to consider the fates of Giordano Bruno and Galileo, the Semmelweis reflex, and all the left-handed people who suffered persecution for centuries.

A number of approaches characterize scientific inquiry. Fundamentally, scientists develop hypotheses, then design and conduct experiments to test them. They analyze and interpret results, summarize their methodology and findings, and publish to inform other scientists. Others, in turn, may reproduce the research to test the validity of findings under different conditions or to examine the effect of modifying assumptions, scope or focus of inquiry, or approach used. Findings are again published to further inform the scientific community. The results of a single study are seldom definitive, with truth emerging over time after repeated experimentation using duplicate and alternative approaches, publication, accumulation of evidence, analysis and reanalysis, debate, and so on.

Occasionally, conflicting scientific findings are cited as evidence that science is useless. What this kind of observation really indicates is a misunderstanding of how science works. In this regard, current use of the term "climate denier" is most unfortunate—and also indicative of misunderstanding of the scientific method. It is important to recognize that Copernicus, Bruno, and Galileo were all, in their time, scorned (and even killed) because they were "the Earth is at the center of the universe" deniers. Some of what we "know" today about climate change may very well turn out to be false.

What is accepted as fact at any point in time is determined by the preponderance of evidence. The door is never closed to new approaches of inquiry that may either further confirm or challenge accepted truth. The system isn't perfect, and not all science is unbiased. Nonetheless, ongoing scientific inquiry and public understanding of scientific methods and of the scientific process are vital in defining and solving current and future problems.

References

Birx, H. 1997. "Giordano Bruno." *Harbinger,* November 11.

Bowyer, J. 1995. "Facts vs. Perception." *Forest Products Journal* 45(11/12): 17–24.

Bowyer, J. 2000. "Tropical Deforestation: Uncovering the Story." *Forest Products Journal* 50(2): 10–18.

Bowyer, J., K. Skog, S. Bratkovich, J. Howe, and K. Fernholz. 2009. *The Wilderness Society Report on Wood Products and Carbon Storage: A Critical Review.* Dovetail Partners, Inc., Sept. 22.

Bush, V. 1945. *Science: The Endless Frontier.* A Report to the President. U.S. Government Printing Office.

Easterbrook, G. 1995. *A Moment on the Earth.* New York: Penguin Books.

Fitzsimmons, J. 2012. "Meet the Climate Denial Machine." *Media Matters,* Nov. 28.

Fumento, M. 1996. *Science under Siege: How the Environmental Misinformation Campaign is Affecting Our Lives.* New York: William Morrow and Company.

Ingerson, A. 2009. *Wood Products and Carbon Storage: Can Increased Production Help Solve the Climate Crisis?* Wilderness Society, April.

Kusher, H. 2011. Retraining the King's Left Hand. *The Lancet* 377(9782): 1998–1999, June 11.

Lomborg, B. 2001. *The Skeptical Environmentalist: Measuring the Real State of the World.* Cambridge University Press.

Marshall Institute. 1997. *Are We Building Environmental Literacy?* A Report of the Independent Commission on Environmental Education. April 15.

Michaels, D. 2008. *Doubt is Their Product.* Oxford University Press.

Moore, P. 2005. "The Environmental Movement: Greens Have Lost Their Way. Scare Tactics, Disinformation Go Too Far." *Miami Herald,* Jan. 28.

Moore, P. 2010. *Confessions of a Greenpeace Dropout: The Making of a Sensible Environmentalist.* Vancouver, B.C.: Beatty St. Publishing, Inc.

Roth, M. 2005. *The Left Stuff: How the Left-Handed Have Survived in a Right-Handed World.* New York: M. Evans.

Rubenstein, B. 1991. "The Fate of the Forest." *City Pages* (Minneapolis), February 27.

Sanera, M. and Shaw, J. 1996. *Facts Not Fear: A Parent's Guide to Teaching Children about the Environment.* Washington, D.C.: Regnery Publishing, Inc.

Semmelweis University. 2016. Ignac Semmelweis Life Story.

Union of Concern Scientists. 1997. "Smoke, Mirrors, and Hot Air: How ExxonMobil Uses Big Tobacco's Tactics to Manufacture Uncertainty on Climate Science."

PART 2

THE CHALLENGE

Former Google CEO Eric Schmid reported in 2010 that every two days as much information is created as was generated from the dawn of civilization up to 2003. The availability of information is staggering. In addition, change is rapid, continuous, and occurs against a backdrop of attempting to balance work and family responsibilities, finances, and all the rest.

Despite, and perhaps because of, the wealth of readily available information about what is going on all around us, it is becoming increasingly difficult to connect the dots. What, for example, are the implications of a 2 percent, 5 percent, or 10 percent annual rate of growth? Why should we care? And how, if at all, do growth trends relate to questions about sustainability? To understand these things requires a basic understanding of the mathematics of growth and concepts of economics. These are the subjects of Part Two.

CHAPTER 4

The Power of Compounding

The greatest shortcoming of the human race is our
inability to understand the exponential function.

—DR. ALVIN BARTLETT

If we are to solve the conundrum of sustainability, we must first come to grips with the power of compounding. The term "compounding" is most often used in the context of interest payments. "Compound interest" is interest paid on a principal sum *and* previously accrued interest. Used more broadly, compounding refers to growth of any quantity, where the rate of growth is applied to a progressively larger number arising from an original value plus periodic increments.

As increasingly sophisticated investors we are well aware that even a small amount of periodic, regular savings can result in a large amount of money over a long period of time, even when interest rates are modest. But as the late Dr. Alvin Bartlett, former professor of physics at the University of Colorado, often pointed out, it seems that we give little thought to the power of compounding when considering the growth of populations, economies, consumption, or anything else outside the realm of banking.

A Mythical Chess Game

Dr. Bartlett often related the mythical story about the inventor of the game of chess. As the story goes, a mathematician had been commissioned by his rather bored king to come up with a new game. In this new game real-life pawns, rooks, and bishops battled their way from square to square in a large checkerboard configured courtyard, all while being directed by the king and his chess opponent who were looking down from the highest battlement of the castle. The king was so appreciative of this new diversion that he asked the mathematician how he could show his gratitude. To this, the mathematician replied that he was merely the king's humble servant. All he asked is that a single grain of wheat be placed on the first square of the chess board, and that for each of the remaining 63 squares, the number of grains be doubled with each successive square.

Courtyard Layout for the King's New Game

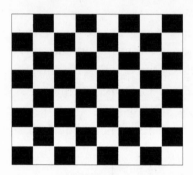

The king, it is said, was bemused by this. He had been prepared to bestow considerable wealth on the mathematician, but received instead only this simple request. He assented with a wave of the hand. However, what the mathematician knew, and that the king did not, was that anything beyond several doublings has an enormous effect upon the resulting number. In fact, doubling any number ten times results in a total that is 1,024 times the original number (i.e., 1-2-4-8-16-32-64-128-256-512-1,024). Looking at this sequence, it is perhaps obvious that doubling 53 additional times would yield an almost unimaginably large number.

It turns out that 63 doublings of that one initial grain of wheat would yield 9.2×10^{18} or 9.2 quintillion grains of wheat, or about 288,099,246,433 metric tons, a quantity of wheat over 426 times larger than the 2012 global wheat harvest! The story does not indicate whether the mathematician survived the king's surprise and then chagrin when he realized the implications of their agreement.

The Realities of Doubling

As illustrated above, repeated doublings quickly add up to very large numbers. However, values don't always double—at least immediately. Sometimes they grow at slower rates. Any rate of growth, if maintained long enough, will eventually result in a doubling of the initial value. For instance, an annual growth rate of 7 percent will result in a doubling of any initial value in 10 years. A simple way to estimate doubling time is to divide the value 70 by the rate of growth. This works when the growth rate is expressed as a percentage. In the case of a 7 percent annual growth rate, the doubling time would be 70 divided by $7.0 = 10$ years. Similarly, the time required to double an investment of \$1 at an interest rate (or annual rate of growth) of 2 percent is approximately 70 divided by $2.0 = 35$ years. Often referred to as the "rule of 70", this simple technique provides a quick way to assess the impact of a given rate of growth. Though not precise, the method yields reasonably close approximations of doubling time.

Understanding compounding has many applications beyond banking and money. Consider, for instance, the following examples:

- The Philippines has a land area $\frac{1}{30}$ that of the United States and a population of about 101 million, just under $\frac{1}{3}$ that of the U.S. It also has an annual population growth rate of 1.7 percent. We are interested in finding out approximately what the population will be two generations into the future (or 40 years from now).

  The fastest way to figure this out is to use a precise mathematical formula. However, using the rule of 70, a quick division (70 divided by the annual growth rate of 1.7 percent) shows that 40 years is almost exactly one doubling time: 70 divided by $1.7 = 41.2$, indicating that at a 1.7 percent rate of growth, the doubling time is 41.2 years. Coincidentally, this is close to the two-generation time span that we are interested in. Based on this simple calculation, it can be quickly estimated that the population of the Philippines will increase within a span of only two generations by almost 100 million if the current growth rate is maintained, bringing the population to almost two-thirds the 2015 population of the United States.

- Aluminum consumption globally grew at an average annual rate of 4.7 percent over the fifteen year period 1995–2010, and is expected to grow 6.5 percent annually over the period 2010–2020 (Alcoa estimate). What do these numbers tell us?

A quick back-of-the-envelope calculation reveals that aluminum consumption doubled in the fifteen-year period 1995–2010, and is expected to nearly double again by 2020: 70 divided by 4.7 = 14.9 years (indicating an approximate doubling of aluminum consumption during the period 1995–2010). Similarly, 70 divided by 6.5 = 10.8 years (indicating a near-doubling from 2010–2021). This also tells us that consumption in 2021 is likely to be roughly four times that of 1995, the result of two doublings.

Small Growth Rates—Large Numbers

Few would be attracted by an interest rate of 0.70 percent, a rate that also happened to be the annual growth rate of the U.S. population in 2016. Nonetheless, earning 0.70 percent with annual compounding would double an investment in exactly 100 years. Doubling would occur in 70 years at an annual growth rate of 1.0 percent. Populations double in precisely the same time frame at this rate of growth. There is, however, a difference between growth of bank accounts and growth of population. Whereas 70 and 100 years are long time periods to an individual, these same time periods represent scarcely a blip in the long sweep of history. This difference helps to explain why people tend to view a 0.70 percent rate of growth as insignificant. But a growth rate of 0.70 percent is far from insignificant.

Keeping in mind that doubling any number ten times results in a total that is 1,024 times the original number, let's examine population growth in the context of compounded annual rates of growth. Remember that a 0.70 percent compounded rate of growth translates to a doubling time of 100 years (70 divided by 0.70 = 100). The population of the U.S. on January 1, 2017 was 324 million. Let's see what it would be in coming centuries with a mere 0.70 percent growth rate.

Using the doubling formula allows a quick estimation at 100-year intervals. As a doubling would occur each 100 years, the increase over a 300-year period would be eight-fold. At the end of 300 years (2317), the U.S. would have a population of 2.6 billion, more than one-third the current population of the entire world. Even maintaining a 0.70 percent growth rate for 200 years, a relatively brief time period in the big scheme of things, would result in a four-fold increase, bringing the U.S. population to about 94 percent of the January 1, 2017 population of China.

Future Population of the United States at an Annual Growth Rate of 0.82 Percent

YEAR	TIME SPAN	POPULATION	POPULATION RELATIVE TO 2017
2017	—	324 million*	—
2117	100 years	648 million	2 x larger
2217	200 years	1.3 billion	4 x larger—about 94 percent of the early 2017 population of China
2317	300 years	2.6 billion	8 x larger—over ⅓ the early 2017 population of the entire world

The impact of a sustained 0.70 percent growth rate on the populations of the various states is shown in the table below. At this rate of growth the combined populations of just three states—California, Texas, and Florida—would be larger in 2216 than the 2016 population of the entire United States. The populations of Florida, Georgia, Illinois, Michigan, New Jersey, New York, North Carolina, Ohio, and Pennsylvania would each be larger than present day California—currently the most populous state.

Population of the 50 States of the U.S. in 2016 and 2216, with Future Growth Assumed 0.70 Percent Annually

STATE	POP. ON JULY 1, 2016 (MILLION)	POP. IN 2216 (MILLION)	STATE	POP. ON JULY 1, 2016 (MILLION)	POP. IN 2216 (MILLION)
Alabama	4.9	19.6	Michigan	9.9	39.6
Alaska	0.7	2.8	Minnesota	5.5	22.0
Arizona	6.9	27.6	Mississippi	3.0	12.0
Arkansas	3.0	12.0	Missouri	6.1	24.4
California	39.3	157.2	Montana	1.0	4.0
Colorado	5.5	22.0	Nebraska	1.9	7.6
Connecticut	3.6	14.4	Nevada	2.9	11.6
Delaware	1.0	4.0	New Hampshire	1.3	5.2
D.C.	0.7	2.8	New Jersey	8.9	35.6
Florida	20.6	82.4	New Mexico	2.1	8.4
Georgia	10.3	41.2	New York	19.7	78.8
Hawaii	1.4	5.6	North Carolina	10.1	40.4
Idaho	1.7	6.8	North Dakota	0.8	3.2
Illinois	12.8	51.2	Ohio	11.6	46.4
Indiana	6.6	26.4	Oklahoma	3.9	15.6
Iowa	3.1	12.4	Oregon	4.1	16.4
Kansas	2.9	11.6	Pennsylvania	12.8	51.2
Kentucky	4.4	17.6	Rhode Island	1.1	4.4
Louisiana	4.7	18.8	South Carolina	5.0	20.00
Maine	1.3	5.2	South Dakota	0.9	3.6
Maryland	6.0	24.0	Tennessee	6.7	26.8
Massachusetts	6.8	27.2	Texas	27.9	111.6

STATE	POP. ON JULY 1, 2016 (MILLION)	POP. IN 2216 (MILLION)	STATE	POP. ON JULY 1, 2016 (MILLION)	POP. IN 2216 (MILLION)
Utah	3.1	12.4	West Virginia	1.8	7.2
Vermont	0.6	2.4	Wisconsin	5.8	23.2
Virginia	8.4	33.6	Wyoming	0.6	2.4
Washington	7.3	29.2	United States	323.1	1,292.4

With the long-term impact of even small annual rates of growth in mind, consider the response, as reported by Dr. Bartlett, of nine members of the Boulder, Colorado City Council in the mid-1960s to the question "What rate of growth of Boulder's population do you think it would be good to have in the coming years?" Responses ranged from a low of 1 percent to a high of 5 percent annually.

Boulder lies at the base of the Rocky Mountains. At an elevation of 5,430 feet, the city enjoys spectacular scenery, and touts its healthy environment and lifestyle. In 1970 Boulder's population was about 67,000. By mid-year 2014 it was about 105,000. Had the population of Boulder grown at a 5 percent rate of growth during that same 44 year period, its population would have been about 617,000 by the end of 2016.

Boulder, Colorado

Photo: Shane Rich for Downtown Boulder.

Boulder Population at a Sustained 5 Percent Annual Growth Rate

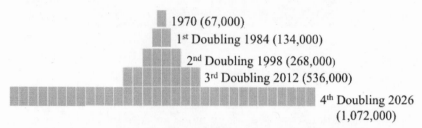

1970 (67,000)

1st Doubling 1984 (134,000)

2nd Doubling 1998 (268,000)

3rd Doubling 2012 (536,000)

4th Doubling 2026
(1,072,000)

* Estimates using the Rule of 70 are approximate. More precise
calculation of population yields larger numbers than shown above.

At a 5 percent annual growth rate, any population will double in size
every 14 years. In the course of 70 years a population growing at this rate
will double 5 times (70 divided by 14 = 5) with the result that the population would be about 32 times larger than in the beginning. What this
could have meant for Boulder is that by 2026 the population would have
exceeded one million, and that by 2040 (or at the 70 year mark following 1970) a 5 percent annual growth rate would have led to a population
of over 2.2 million! Depending upon your perspective, a population of
617,000 or 2.2 million might be viewed as either good or bad. But there
is no doubt that the character of Boulder would be vastly different under a
5 percent annual growth rate than at a lower rate of growth. It is important
to recognize, however, that at *any* sustained rate of growth, Boulder's
population will eventually reach 2.2 million and beyond.

Beyond the issue of population growth, the concept of compounding applies to growth of any quantity. Earlier, compounding calculations
were used to determine the magnitude of growth in aluminum consumption over a period of time. Consider also how compounding helps to com-
prehend consumption trends of other vital resources:

- Global grain consumption doubled between 1950 and 1976 then doubled again between 1976 and 2016, resulting in a quadrupling of grain
  consumption over a 66-year period.
- Fresh water withdrawals globally increased by a factor of 2.8x between

1950 and 2016, and are growing at 1.6 percent annually—a 44-year doubling time.

- Primary energy consumption globally doubled during the period 1957–1972 and doubled again by 2015—a four-fold increase over a period of 58 years. The annual growth rate as of 2016 was 1.5 percent, translating to a 47 year doubling time.

As with population, even seemingly minor rates of growth of consumption can result in large increases in amounts consumed.

The impact of successive doublings on numbers is the same regardless of the rate of growth. For instance, the previous example showing that any number will be 32 times larger after five doublings holds true whether the rate of growth is 5, 3, or 1 percent or at any other rate, and also regardless of whether the growth rate is steady or variable.

An inescapable conclusion is that population growth cannot continue indefinitely. It is also clear that growth rates often viewed as negligible or modest can lead to unmanageable numbers within a relatively short period of time. Even a rudimentary understanding of compounding can provide insights that are otherwise not obvious. Such understanding is critical to appreciation of fundamental trends underlying a wide array of environmental impacts.

Exponential Growth

References are sometimes made to *exponential* growth, a situation in which growth occurs at a fixed percentage at regular intervals, as in all of the examples provided to this point. At each compounding interval a number becomes larger than it previously was. Consequently, the size of increase becomes progressively greater with each compounding period, even though the percentage rate of increase remains constant.

Consider the following example of exponential growth:

- Starting with the number 100 and a 10 percent annual rate of increase, the first compounding results in an increase of 10 (100 x 1.1 = 110).

- After nine more compounding periods the original 100 number will have grown to 258. Growth in the next year will add another 10 percent. However, instead of an increase of 10 as in year 1, compounding by the same 10 percent annual rate yields an increase of 26 in year 11.
- By year 20 the original number will have grown to 736, with the result that the increase in year 21 is 74. And so on.

A constant rate of increase leads to greater and greater absolute increases in any number, beginning at any starting point. This shows up when this kind of growth sequence is plotted graphically below, clearly showing acceleration of growth in numbers over time.

Value of Initial Value of 100 Compounded at 10 Percent per Time Period

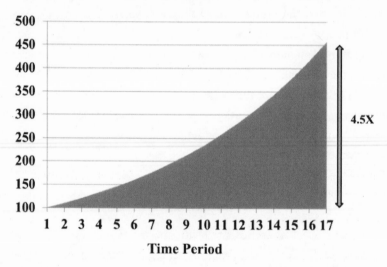

Accelerated growth also is evident in a plot of world population growth, the result of exponential growth over many centuries.

World Population 10000BC–2000AD

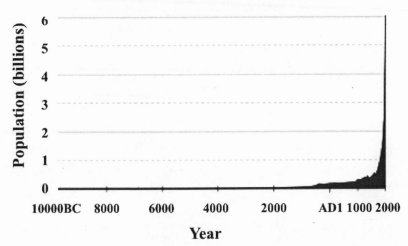

This graphic has been released into the public domain by its author, El T at English Wikipedia

Although not evident in the previous graphic, the annual percentage rate of world population growth began to slow in the early 1960s. Since that time the rate of growth has continued to decline slowly, meaning that growth is no longer exponential, even though many people continue to refer to "exponential" population growth. While this may be good news, a declining rate of growth does not automatically translate to declining numbers of people being added to the population. As discussed in the following Chapter, applying a slightly smaller rate of growth to a bigger and bigger population can still add large numbers each year to an already very large population.

An interesting aspect of repeated doubling is that the sum of all of the numbers up to and including the most recent doubling are greater than for all previous doublings combined. For purposes of illustration, assume a supply of a certain non-renewable material of 300 million tons. Further assume that consumption in the first time period is 1 million tons, with consumption doubling with each tick of the clock. Consumption would proceed: 1, 2, 4, 8, so that after four ticks of the clock, total consumption to that point would be 15 million tons ($1 + 2 + 4 + 8 = 15$). But at the next tick of the clock consumption will be 16 million tons (2 x consumption in the previous period, or 2 x 8), a number greater than all previous consumption combined.

Successive Doubling of Consumption Illustrating that Each Doubling Results in Consumption Greater than All Previous Consumption Combined

CONSUMPTION IN EACH SUCCESSIVE TIME PERIOD (MILLION TONS)	CUMULATIVE CONSUMPTION AT END OF TIME PERIOD (MILLION TONS)
1	1
2	3 (1 + 2)
4	7 (3 + 4)
8	15 (7 + 8)
16	31 (15+16)
32	

In a situation in which growth of a non-renewable resource is exponential, as is the case with global growth in petroleum consumption, the impact of doubling on historic consumption becomes especially relevant. Consider a situation in which it is announced that remaining reserves are equal to total consumption over all of history. This is surely a reassuring statement. Or is it? What is really being said in this case is that reserves are likely to be exhausted in the course of the next doubling, or within one doubling time.

A Seemingly Easy Choice

The point of this discussion has been to create an appreciation of the power of compounding. Over any appreciable length of time the impact can be enormous even when the rate of increase is seemingly negligible.

In case the point was missed, or to give you a bit of ammunition to mess with the minds of the people around you, consider the following example, provided by Wealth Informatics, that is similar to the one we began with, but tantalizingly different:

Having won the grand prize in a state fair contest you are offered a delightful choice:

- $1,000,000 in cash
- 1¢ which is deposited into an account that doubles in value each day for 31 days.

Intuition would perhaps suggest that taking the one million dollar cash prize is the clear choice. Doing a few calculations regarding the other choice would perhaps confirm this. By the end of day 10, anyone choosing the second option would have in their account only $5.12, and by the end of the 20th day only $5,242.88. However, with only 8 more doublings, and on day 28, the value of the original 1¢ account would now be $1,342,177.28. But, there would be three more days still to go. At the end of the 31 days, the account that began with deposit of a single penny would now be worth $10,737,418.24! The same amount would result from investing one cent in an account with a 10 percent annual interest rate and leaving it there for a little over 218 years, or investing that same one cent in an account with a 3 percent annual interest rate and leaving it there for a bit over 703 years. Once again, the point is that any rate of growth extended for a long period of time can make a seemingly modest or manageable number into a very, very large number.

References

Bartlett, A. [n.d.] Arithmetic, Population, and Energy. Video and written transcript. Accessed April 10, 2014.

Wealth Informatics. 2014. Double a Penny a Day for 31 Days. Accessed April 10, 2014.

U.S. Census Bureau. 2015. U.S. and World Population Clocks.

CHAPTER 5

Growth of the Masses

Somewhere on this globe, every ten seconds,
is a woman giving birth to a child.
She must be found and stopped.

—SAM LEVENSON

Today, it is common to hear references to billions—billions of dollars, billions of hamburgers, billions of people . . . But have you ever thought about just how many a billion is? Consider these calculations made in February 2017:

- A billion seconds ago Ronald Reagan was beginning the fourth year of his presidency and the first IBM personal computer would not be on the market for several months.
- A billion minutes ago a Jewish revolt against the Roman Empire was close at hand, and China was ruled by the Han Dynasty.
- A billion hours ago Neanderthal man inhabited parts of Asia, Europe, and Africa, and the first crude stone tools would not be used for another 100,000 years.
- A billion days ago mammoths and mastodons lived on the North American plains.

Large and Growing Numbers

The first modern humans (*Homo sapiens*) are known to have walked on Earth at least 200,000 years ago. Scattered populations grew slowly at first, and as recently as 1,000 BC our species numbered no more than

200 to 300 million. Implementation of basic sanitation and gradual improvement of farm practices allowed the population to increase, and by 1800 humans numbered about 1 billion. One hundred thirty more years would elapse before the two billion mark was reached (1930). Modern medicine would then begin to impact the far corners of the Earth, and with stunning results. The global population hit 3 billion in 1960 (a 30-year time span), 4 billion in 1974 (14 years), 5 billion in 1987 (13 years), 6 billion in early 1999 (11½ years), and 7 billion in 2012 (13 years). In early 2017 the population was about 7.4 billion.

In mid-2016 the global population was increasing at a rate of about 213,000 each day, more than 8,860 each hour. To put this rate of increase into perspective, check your watch and begin counting out loud as fast as you can for one minute. Ready? Go . . .

. . . so how did you do? If you counted to anything less than 148 your counting, as fast as it may have seemed, was slower than the rate of increase in global population (births minus deaths). You would have to count to 148 every minute of every day to match the population growth rate. An incredibly fast counter might have reached 258 in the span of one minute. In this case the rate of counting would equal the rate of births globally!

Growth Is Rapid—But At A Declining Rate

As spectacular as the growth rate has been, it is important to realize that the rate has actually been falling for about 80 years. Birth rates have been falling steadily worldwide since the mid-1930s. Death rates have fallen as well, but generally less rapidly. These two realities in combination translate to declining growth rates in both absolute numbers and percentage.

Global Rates of Birth, Death, and Natural Increase per 1,000 People

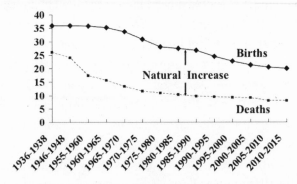

Source: Data for 1936–2000 from United Nations, *World Population Prospects: The 2002 Revision.*
Data for 2001–2005 from Population Reference Bureau, 2015.

In the previous chapter, population growth was shown graphically over thousands of years of human history. The rate of growth over the past 65 years was stunning. But when examined over a shorter time scale, a slowing of population growth in recent decades is obvious. Thus, while the population is growing very rapidly, it is not growing, as some continue to suggest, at an exponential rate.

World Population, 1850–2050

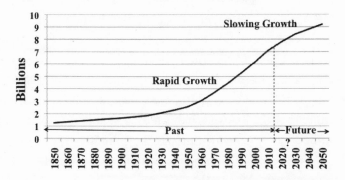

Source: U.S. Census Bureau, International Programs Center, 2015.

That population is increasing, but at a declining rate, is a source of a great deal of confusion. How does news of a declining rate of growth square with the previous discussion of rapidly growing populations?

Some interpret news of a declining growth rate to mean that the population is not growing. Others have been inspired to declare that population is no longer an issue that humans need to be concerned about.

Both of these views reflect a misunderstanding of population trends. This is because a decline in the rate of growth does not equate to a decline of growth in numbers. While the *rate* of growth has dropped, growth in numbers has not. For the past 80 years the rate of population growth has been falling at the same time that annual increases in population numbers have been rising. Moreover, as discussed in Chapter 4, even a seemingly very small rate of increase, if applied to a large population, can translate to absolute growth on a very large scale.

The good news in all of this, as pointed out by Swedish medical doctor and statistician Hans Rosling and UN population statistics, is that the number of children in the world is today growing very slowly, suggesting population stabilization in the not-too-distant future. Overall population trends indicate the same thing: that population numbers are likely to peak (i.e., reach a point at which the population will level off or begin to decrease in size) within the next century or so. The most recent (2015) world population projection of the United Nations Population Division forecasts a population of 9.7 billion by 2050 and 11.2 billion by 2100, with an 80 percent chance that the population in 2100 will be between 9.6 and 13.3 billion. Using different forecasting methods, the International Institute for Applied Systems Analysis (IIASA) indicated in 2014 that world population would likely reach a peak in this century. According to IIASA, numbers are likely to increase to 9.2 billion by 2050, peak at 9.4 billion around 2070, and then start a slow decline to 9.0 billion by the end of the century. Both forecasts indicate a need to find a way to accommodate 2 billion more people within just the next 50 years (a significant challenge by any measure), and perhaps as many as 6 billion or more for the very long term. Among the challenges is finding ways to feed increased numbers. The UN Food and Agricultural Organization estimates that world agricultural output will need to grow by 70 percent over 2010 levels to feed a population of 9 billion with rising levels of consumption.

Wild cards in population forecasting are the uncertainty inherent in all predictions of future events, and the extent to which current trends might change. That forecasts can change, and substantially, is evidenced by upward revision of UN population projections in 2015, with new numbers indicating 9.7 billion by 2050 rather than 8.9 billion as predicted earlier. In this case, undercounting of children and fertility rates in Africa, and increasing life expectancy worldwide were cited as reasons for higher projected numbers. The upward revision was made prior to China's abandonment of its one-child policy. Future shifts in population trends could be driven by rising concern over prospects of population stabilization or decline. As of this writing, initiatives to increase the rate of childbirths are underway in a number of countries, including Japan, Russia, Turkey, Singapore, and a number of member states of the European Union.

Population Trends in the United States

The Past as Prologue to the Future

On October 17, 2005, U.S. population numbers officially reached the 300 million mark. Images from that day include a group of people clustered around a population clock somewhere in the nation's capital who applauded as the number surpassed 300 million, and a *New York Times* editorial stating that the 300 millionth person should receive a bouquet and a thank-you card from President Bush. A happy day indeed!

Trends in U.S. population growth are significantly different than growth globally. There is no obvious sustained downturn in the growth rate in the historical record, although population projections suggest a steady slowing of growth rate going forward. Nonetheless, the projected 2060 population is more than double the actual population in 1970.

Growth of U.S. Population, 1776–2010

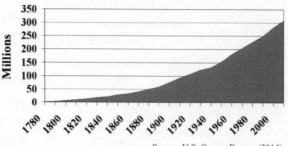

Source: U.S. Census Bureau (2014).

From the mid-1970s through about 2006 the U.S. national population growth rate mostly hovered between 0.9 and 1.0 percent annually. The birth rate slowed markedly during the recession of 2007–2010, and continues to lag the earlier rate. In early 2016, growth of the U.S. population was driven by immigration and a birth rate that exceeded the death rate. Every two minutes, growth was defined by the arrival of 4 new immigrants, and 15 births and 9 deaths, for a net gain of 10.

So What?

The impact of population growth on the availability of fixed assets such as land, water, and forests on a per capita basis is both relentless and subtle. As one example, consider the area of forest land in the United States at different points in history. In 1785 there were about 350 acres of forest for each man, woman, and child. This declined dramatically over a period of only 65 years due to a combination of conversion of forest lands to agriculture (resulting in a loss of more than 11 percent of forest land) and a more than 7-fold (or more than 676 percent) growth in population. By 1850 the area of forest was 40 acres per capita. Similar developments occurred over the next 60 years, reducing the forest land area per capita to 9.5 acres—only $\frac{1}{35}$ of what it had been 125 years earlier.

U.S. Forests Then, Now, and Future

YEAR	POPULATION (MILLION)	FOREST AREA (MILLION ACRES)	FOREST AREA/CAPITA (ACRES)
1785	3.0	1,044	348
1850	23.3	926	40
1910	77.0	730	9.5
2016	323.1	766	2.4
2100	571.0	766	1.3

Source: Population figures from U.S. Census Bureau (2015);
forest area from USDA-Forest Service.

Over the succeeding 102 years the area of forest land in the U.S. actually increased as abandoned farmland reverted back into forest cover. However, the area of forest on a per capita basis continued to drop as the population grew from 77 to 323 million. Assuming a stable forest land base going forward, forest area will continue to fall on a per capita basis, with slightly over 1.3 acres the most likely value by the end of the 21st century.

Similar per capita trends can be seen in the volume of fresh water, the extent of lake shore, the area of tillable land, the quantity of mineral resources within our borders, and so on. Growth of human numbers also directly and indirectly impacts species other than humans. Writing in the *Wall Street Journal,* Heidi Vogt noted that despite the outcry over the killing of Cecil the lion in 2015, human population growth is a larger threat by far to the future of Africa's lions.

Beyond the reality of falling per capita availability of fixed assets, growth in numbers results inevitably in consumption growth. As related by Thomas Friedman in his book *Hot, Flat and Crowded,* the effect of providing a single lightbulb to a billion additional people would result in added resource consumption approximating 20,000 metric tons just to make the bulbs and their packaging. Additional resources would be needed to make the lamps needed to screw the bulbs into. Turning the bulbs on would create additional resource needs. To light those billion bulbs the equivalent of twenty new 500 megawatt coal-burning-power plants would be needed even if they were burned just four hours each day.

Population matters. As the population continues to grow, the per capita quantity of fixed resources, such as land, becomes incrementally smaller, and the margin for error in meeting human needs narrows steadily.

Population Growth from the Perspective of a U.S. President

The U.S. president that did more than any other to call attention to negative implications of population growth is Richard Nixon. Doubtless influenced by publication of Paul Ehrlich's *The Population Bomb* only a year earlier, Nixon in July 1969 released a *Special Message to the Congress on Problems of Population Growth.* Mr. Nixon, of course, is remembered almost totally in the context of the Watergate scandal and the events that followed, and it is those events, and the simmering stew of politics of a re-election year (in reverse order) that completely derailed the president's population initiatives. Nonetheless, his words of 45 years ago are worth reading, if for no other reason than to contemplate how present challenges have been shaped by population growth. In a report to Congress, the President began by noting that:

> . . . *Most informed observers . . . agree that population growth is among the most important issues we face. They agree that it* [the challenge] *can be met only if there is a great deal of advance planning. And they agree that the time for such planning is growing very short."*

He went on to say that

> *"If the present rate of growth continues, the third hundred million persons will be added in roughly a thirty-year period. This means that by the year 2000, or shortly thereafter, there will be more than 300 million Americans.* [There were 309 million Americans by mid-2010 and 324 million by mid-2016].

Nixon continued:

This growth will produce serious challenges for our society . . . Where, for example, will the next hundred million Americans live? If the patterns of the last few decades hold for the rest of the century, then at least three quarters of the next hundred million persons will locate in highly urbanized areas. Are our cities prepared for such an influx? The chaotic history of urban growth suggests that they are not and that many of their existing problems will be severely aggravated by a dramatic increase in numbers. Are there ways, then, of readying our cities? Alternatively, can the trend toward greater concentrations of population be reversed? Is it a desirable thing, for example, that half of all the counties in the United States actually lost population in the 1950's, despite the growing number of inhabitants in the country as a whole? Are there ways of fostering a better distribution of the growing population? . . . If we were to accommodate the full 100 million persons in new communities, we would have to build a new city of 250,000 persons each month from now until the end of the century. That means constructing a city the size of Tulsa, Dayton, or Jersey City every thirty days for over thirty years. Clearly, the problem is enormous, and we must examine the alternative solutions very carefully . . .

. . . What of our natural resources and the quality of our environment? Pure air and water are fundamental to life itself. Parks, recreational facilities, and an attractive countryside are essential to our emotional well-being. Plant and animal and mineral resources are also vital. A growing population will increase the demand for such resources. But in many cases their supply will not be increased and may even be endangered. The ecological system upon which we now depend may seriously deteriorate if our efforts to conserve and enhance the environment do not match the growth of the population . . .

. . . Finally we must ask: how can we better assist American families so that they will have no more children than they wish to have? In my first message to Congress on domestic affairs, I called for a national commitment to provide a healthful and stim-

ulating environment for all children during their first five years of life. One of the ways in which we can promote that goal is to provide assistance for more parents in effectively planning their families. We know that involuntary childbearing often results in poor physical and emotional health for all members of the family. It is one of the factors which contribute to our distressingly high infant mortality rate, the unacceptable level of malnutrition, and the disappointing performance of some children in our schools. Unwanted or untimely childbearing is one of several forces which are driving many families into poverty or keeping them in that condition. Its threat helps to produce the dangerous incidence of illegal abortion. And finally, or course, it needlessly adds to the burdens placed on all our resources by increasing population ...

... Perhaps the most dangerous element in the present situation is the fact that so few people are examining these questions from the viewpoint of the whole society. Perceptive businessmen project the demand for their products many years into the future by studying population trends. Other private institutions develop sophisticated planning mechanisms which allow them to account for rapidly changing conditions. In the governmental sphere, however, there is virtually no machinery through which we can develop a detailed understanding of demographic changes and bring that understanding to bear on public policy. The federal government makes only a minimal effort in this area. The efforts of state and local governments are also inadequate. Most importantly, the planning which does take place at some levels is poorly understood and is often based on unexamined assumptions.

Subsequent to Nixon's missive, he formed, in 1970, a National Commission on Population and the American Future and appointed most of its members, including Commission chairman John Rockefeller. After several years of work, the Commission responded with 70 recommendations; key among them were that legal immigration be frozen at 400,000 annually, that illegal immigration be stopped, and that sex education be made widely available in schools. One recommendation, however, proved to be the undoing of the entire effort: a call for decriminalization of abortion. Facing what was anticipated to be a difficult re-election campaign, and stiff opposition from religious leaders and the right wing of the party to the Population Commission's report, and to the abortion provision in particular, Nixon rejected the report's recommendations, strongly condemning the abortion aspect.

Despite the fallout from the Rockefeller Population Commission report, Nixon took steps during his tenure to extend financial and technical assistance to developing nations for population planning programs, to match population planning-related contributions to the United Nations from other countries, and to support through appropriate governmental agencies private decisions of American citizens that could slow domestic population growth. Under his watch, federal funding for family planning services within the United States was initiated (1970) as Title X of the Public Health Service Act, and Medicaid funding for family planning was authorized (1972).

Under subsequent presidents, support for population initiatives shifted dramatically from administration to administration, similar to the flight of a ping pong ball in a highly-contested tournament. The U.S. was, for example, the largest contributor to the United Nations Population Fund (UNFPA) from 1969, Nixon's first year in office, through 1985 when President Reagan halted funding in response to reports of a link between population program funding and human rights abuses in China. U.S. funding of UNFPA remained on hold through the term of President H.W. Bush, was restored as one of the first official acts of President Clinton, was again put on hold through the administration of President G.W. Bush, and was restored again within several days following the inauguration of President Barack Obama, and rescinded again on the first day of the Trump administration.

The history of support for voluntary family planning through the U.S. Agency for International Development (USAID) has been a bit less tenuous. Funding for USAID's Office of Population and Reproductive Health was initiated in 1965 by President Lyndon Johnson and has continued through successive administrations at varied levels through the present. Programs in some 65 countries are currently supported. On the domestic front, funding for family planning through the Public Health Service Title

X program, though modest, has endured through every administration since that of Nixon. Support for education of girls and women has also endured through successive administrations, based on recognition that access to education is a fundamental human right. Coincidentally, education of women is also widely recognized as a necessary prerequisite to stemming growth of human numbers.

Sharply Opposing Views

Nixon's views are not, of course, shared by everyone. In 1981, then University of Maryland professor Julian Simon authored *The Ultimate Resource*. In it, he argued that increasing human needs for resources simply triggers learning and innovation and a permanently greater ability to obtain resources. He went on to say that "there is no meaningful physical limit—even the commonly mentioned weight of the Earth—to our capacity to keep growing forever." Dr. Simon also expressed the view that each new child represents brainpower that might help to find solutions to the many challenges faced by mankind. The result was an open feud with *Population Bomb* author Paul Ehrlich, a bet with Ehrlich that future minerals prices would go down over the coming decade rather than up (which Simon won), and considerable flak from the environmental community in general. Nonetheless, many in society found resonance in Simon's message. A dozen years later a senior editor at the Cato Institute authored an article "Much Ado About Nothing: Population Growth as Promise Not Problem." His premise was much the same as that of Simon. The passage of another dozen years found a continuation of the same thread, this time in a staff editorial in the *Wall Street Journal* (October 21, 2006) that said in part ". . . people don't just consume things. They make them too. More bodies mean more minds, more innovation, more dynamism, and more progress."

Others view the merits of growth from an economic perspective. In 2005, on the October commemoration of the 300 millionth American, then Secretary of Commerce Carlos Gutierrez issued a statement to the effect that America's growing population is good for the economy, and necessary to enable the support of aging populations. More recently an article "U.S. Population Growth Slowed, Still Envied" appeared in *USA Today*. The focus of that article was the negative impact of an aging population on financial support for Social Security and Medicare, with all of those interviewed supporting growth as a solution to financing of social programs. High levels of immigration were deemed essential to maintaining population growth.

The Immigration Issue

A Hot Potato

Nothing in the discussion that follows should be interpreted to mean that immigration or immigrants are evil, that nations shouldn't respond to humanitarian crises, or that there aren't myriad benefits of blending races and cultures. It is also recognized that growth of a national population as a result of immigration is different from population growth in a world sense in that immigration does not directly affect the net rate of population growth globally. What is pointed out herein is the reality that immigration can have large environmental implications for the country of destination, and for a very long time thereafter.

Quite apart from consideration of population growth and size, immigration has become a hot issue in recent years in the United States based on a host of other considerations. To some, efforts to limit immigration are an affront to human rights and social equity. Others point to immigration as essential to providing a source of workers to fill undesirable jobs or to support increasing numbers of retirees. In sharp contrast to these views, others believe that controlling immigration is vital to protection of domestic jobs and national security. Curiously, largely missing from public commentary or debate is consideration of environmental impacts of immigration. The topic was touched upon by President Clinton's Council on Sustainable Development, Population and Consumption Task Force in 1996. They wrote "This is a sensitive issue, but reducing immigration levels is a necessary part of population stabilization and the drive toward sustainability."

Over the past decade immigration has accounted for about 30 percent of U.S. population growth, with legal immigration alone accounting for about 850,000 to a million new arrivals annually, a level more than twice that recommended as an upper limit by the Rockefeller population commission. In addition, illegal or unauthorized immigration is estimated to have accounted for another 850,000 new arrivals annually for the period 2000 through 2005, about 550,000 annually 2005 through 2007, and lesser numbers (300,000) through the recession years as economic prospects dimmed. In 2016, total immigration numbers rebounded to near pre-recession levels. The magnitude of the numbers tells an unmistakable story: immigration, legal and illegal, is clearly a significant factor in U.S. population growth.

Immigration: Good or Bad?

The question of whether net international migration is needed to help provide for the large number of people expected to retire over the next several decades has been extensively researched, most notably by the United Nations Population Division. Using statistics from eight geographic regions, including the United States, the UN Population Division examined the question of whether what they termed "replacement migration" is a solution to declining population size, declines in the population of working age, and overall aging of a population.

The UN study considered six scenarios for the United States:

1. A scenario consistent with the medium projection for U.S. population growth. This scenario assumed an annual net influx of 760,000 immigrants from 1995 through 2050, with a rapid rise in the 65 and older population. The support ratio of working adults to retirees (the number of working adults divided by the number of retirees) was projected to decline from 5.2 to 2.8 in 2050. Total projected U.S. population in 2050: 349 million.

2. Assumed medium growth, but with zero net migration. The result here was a projected 2050 population of 290 million and a support ratio of 2.6.

3. Assumed a level of immigration sufficient to keep the U.S. population

constant at the 1995 level; this translated to zero net immigration for the period 1995–2030, and 6.4 million net immigrants annually during the period 2030 through 2050. Under this scenario the projected 2050 population was 298 million and the support ratio 2.6.

4. Assumed a level of immigration sufficient to keep the size of the age 15–64 population constant at 1995 levels; this translated to zero net immigration for the period 1995–2015, and 513,000 net immigrants annually during the period 2015 through 2050. Under this scenario the projected 2050 population was 316 million and the support ratio 2.7.

5. Under this scenario the support ratio was not allowed to drop below 3.0. To achieve this, zero net immigration was needed through 2025, with 4.5 million net immigrants needed annually between 2025 and 2035. In this case the 2050 population was projected at 352 million and the support ratio was 3.04.

6. Assumed in this scenario was a net immigration sufficient to maintain the same support ratio as in 1995 (5.2). Under this scenario the net immigration need was determined to be 10.8 million per year, leading to a 2050 U.S. population of 1.1 *billion,* of which 73 percent would be post-1995 immigrants and their descendants.

United States Net Immigration, Population, and Support Ratio under Six Scenarios

	NUMBER OF IMMIGRANTS ANNUALLY (NET)	2050 POPULATION (MILLIONS)	SUPPORT RATIO
Scenario 1	760 thousand	349	2.8
Scenario 2	Zero	290	2.6
Scenario 3	Zero 1995–2030; 6.4 million annually 2030–2050	298	2.6
Scenario 4	Zero 1995–2015; 513 thousand annually 2015–2050	316	2.7
Scenario 5	Zero 1995–2025; 4.5 million annually 2025–2035	352	3.0
Scenario 6	10.8 million	1,100	5.2

Source: United Nations (2000).

The UN report noted that the recent rate of inflow of immigrants into the United States was presently well above that needed to prevent a decline in the population or in the working-age population. Also noted, however, was that attempting to maintain the 1995 support ratio would involve massive net immigration or, alternatively, an increase in the retirement age to 72. Maintaining a support ratio of 3.0 would be attained, the report concluded, by simply raising the retirement age to 67.

So, regarding the answer to the question of whether significant net immigration is necessary in order to support large numbers of retirees, the answer appears to be at least a qualified "yes" . . . at least in the near term. As stated in the UN report summary "If retirement ages remain essentially where they are today, increasing the size of the working-age population through international migration is the only option in the short to medium term to reduce declines in the potential support ratio." There is, however, a cost to this strategy, and that is reflected in substantial population increases that cannot continue indefinitely.

Since the UN study considered only the future through 2050, a time not all that far away, consideration of what might happen thereafter is in order. Does society, for example, need large numbers of new arrivals for the next 36 years only, after which time the need becomes smaller? For the next 100 years . . . the next 200?

There is, in fact, no obvious end-point for these arguments. A number of investigators have looked at this and have come to a more or less uniform conclusion that over a long period of time, immigration is no solution to an aging population. And it is not a solution because immigrants also age. Chasing larger and larger population numbers to solve a problem that society would rather not currently confront will, in the end, simply result in a much larger population. That larger population will at some point have to face the same problems that confront society today, only the problems at that point will be larger.

Potential Population Decline

What might be our reaction to population decline? Remember that projections suggest peaking of world population at some point, and maybe within this century, with stabilization or perhaps decline thereafter. Because such decline would ultimately affect the U.S. population, it is interesting to see the response to a dip in U.S. population growth brought about by the deep economic recession of 2007–2010—an event that had led to reductions in both immigration and birth rates. Consider, for instance, comments that appeared on the Brookings Institution website a week following the U.S. Census Bureau's release of state population counts for 2011. The commentary described as "disheartening" the reported population growth rate of 0.73 percent, noting that the U.S. population increase for the previous decade had been the lowest since the Great Depression. The author expressed worry about the trend toward slower growth, citing needs for maintaining the largest possible workforce as a percent of the population in order to retain high and rising standards of living, and replenishing the ranks of younger workers in order to bolster the workforce and enable ongoing support for aging citizens. A concluding remark indicated that societal confidence in both the present and future is reflected in population growth, apparently suggesting lack of confidence if the population is not growing.

A more recent article took the hand-wringing to an even higher level, remarking that the IIASA projection that the world population would likely peak, perhaps within the 21st century, could eventually result in "the literal extinction of humanity." With this kind of angst being expressed over slowing growth rates, a relevant question is how future generations are going to handle much steeper reductions in the rate of growth, reductions that at some point must occur. Keep in mind that the impact of population growth on the availability of fixed assets such as land, fresh water, and forests on a per capita basis is both relentless and subtle.

Reflection

An often overlooked finding of the 1972 Rockefeller report, that even business representatives on the Commission reportedly supported, was the central, near heretical finding that "the health of our country does not depend on population growth, nor does the vitality of business, nor the welfare of the average person." We are long overdue in revisiting that conclusion.

References

Anadolu Agency. 2015. "Incentives to Boost Birth Rates to Cost Turkey $400 Min." January 12. Accessed December 15, 2015.

Brainerd, E. 2014. "Can Government Policies Reverse Undesirable Declines in Fertility?" *World of Labor,* May.

Browne, A. 2002. *Do We Need Mass Immigration?* London: Institute for the Study of Civil Society.

Buckley, C. 2015. "China Ends One-Child Policy, Allowing Families Two Children." *New York Times,* Oct. 29.

Ehrlich, P. 1968. *The Population Bomb.* New York: Sierra Club/Ballantine Books.

Friedman, T. 2008. *Hot, Flat and Crowded: Why we Need a Green Revolution and How it Can Renew America.* New York: Farrar, Straus and Giroux.

Frey, W. 2011. "2011 Puts the Brakes on US Population Growth." Brookings Institution, December 28.

International Institute for Applied Systems Analysis (IIASA). 2007. *Population Projections: The 2007 update to IIASA's probabilistic world population projections.* Accessed April 10, 2014.

Lutz, W., W. Betz, and K. Samir. 2014. *World Population and Human Capital in the Twenty-First Century.* Oxford University Press.

McNamee, W. 2011. "US Population Growth Slowed, Still Envied." *USA Today,* Jan. 27.

Mumford, S. 1993. *Overcoming Population: The Rise and Fall of American Political Will.* Center for Research on Population and Security.

Nixon, R. 1969. *A Special Message to the Congress on Problems of Population Growth.* The White House, July 18.

Noack, R. 2015. "Please Make More Babies, Europe Urges its Residents." *Minneapolis Star Tribune,* April 11.

Passel, J., and J. Cohn. 2010. *U.S. Unauthorized Immigration Flows are Down Sharply Since Mid-Decade.* Pew Research Hispanic Center.

Population Reference Bureau. 2015. *World Population Data Sheet.*

President's Council on Sustainable Development. 1996. *Report of Population and Consumption Task Force.*

Richman, S. 1993. "Much Ado about Nothing: Population Growth as Promise, Not Problem." *The World & I* 8(6): 374–375.

Rockefeller Commission. 1972. *Population and the American Future.* Commission on Population Growth and the American Future.

Rosling, H. 2013. *Don't Panic.* Wingspan Productions. Accessed November 10, 2014.

Singapore Government. 2015. "The Baby Bonus Scheme," last modified June 15, 2015. Accessed December 15, 2015.

Simcox, D. 1998. "Nixon's 1969 Special Message to Congress: Anniversary of a Missed Opportunity." *Negative Population Growth, Forum Series.*

Simon, J. 1981. *The Ultimate Resource.* Princeton, New Jersey: Princeton University Press.

United Nations. 2000. *Replacement Migration: Is it a Solution to Declining and Ageing Populations?* Population Division, Department of Economic and Social Affairs, United Nations Secretariat.

United Nations. 2015. *World Population Prospects: The 2015 Revision.* Department of Economic and Social Affairs, Population Division.

United Nations Population Fund. 2011. *Population Matters for Sustainable Development.* Interagency Consultation on Population and Sustainable Development.

U.S. Census Bureau. 2005. "Chart of US Population, 1790–2000." Accessed November 12, 2015.

U.S. Census Bureau. 2015. "US and World Population Clocks."

USDA-Forest Service. 2009. *Forest Resources of the United States, 2007.*

Vogt, H. 2015. "Human-Population Boom Remains Largest Threat to Africa's Lions in Wake of Cecil's Killing." *Wall Street Journal,* August 7.

Wise, J. 2013. "About that Overpopulation Problem: Research suggests we may actually face a declining world population in the coming years." *Slate.com,* January 9.

CHAPTER 6

Economic Miracles

The future isn't what it used to be.

—PAUL VALERY

In 1975 the heads of state of the six democratic countries with the world's largest economies—France, Germany, Italy, Japan, the United Kingdom, and the United States—began meeting annually to discuss major issues of mutual or global concern. Known as the Group of Six (or G-6), the group expanded within a year to include Canada, coming then to be known as the G-7 nations (and later the G-8 with the addition of Russia). Motivated by a sense of responsibility as well as self-interest, and perhaps even a bit of arrogance, G-7 leaders met to deal with a wide range of issues, and sometimes offer advice to leaders of other nations.

When the G-7 was first established, five of its members—France, Germany, Italy, Japan, and the United Kingdom had ranked among the world's economic elite for more than 500 years. Only the United States and Canada were relative newcomers to the list, with the U.S. having moved into a position of prominence only about 150 years before. Interestingly, the real economic power 500 (and even 1,500) years ago resided within China, India, Mongolia, and Persia. In fact, as recently as 1820, the gross national products of Asian nations, including China and India but excluding Japan, were substantially larger than the whole of Europe, Russia, Japan and the Americas *combined*.

The industrial revolution and the resulting economic revolution changed everything. A prolonged period of economic growth for those countries that adopted mechanization and new forms of energy led to a rapid reordering of global wealth, and resulted in the emergence of the American economic juggernaut over a period of less than a century.

The position of the U.S. as an economic and military superpower appears to be taken more or less for granted by Americans today. Moreover, the behavior of those in positions of power and influence suggests a belief that the current situation is likely to persist over the long term. But another revolution is already underway.

Growth of the Global Economy

As the world population continues to grow, the global economy is growing as well, albeit at a much more rapid rate. For instance, between 1970 and 2012, a period in which world population increased 90 percent, the global economy grew 260 percent in inflation-adjusted dollars. Furthermore, economic growth in recent decades has been most rapid in the world's developing nations, nations that for centuries were characterized by poor, agrarian economies.

Gross World Product, 1970–2015

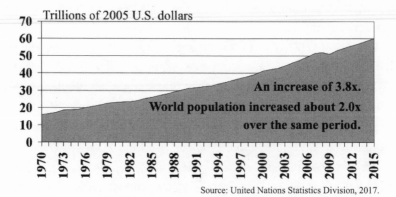

An increase of 3.8x. World population increased about 2.0x over the same period.

Source: United Nations Statistics Division, 2017.

Economies worldwide have long been in a growth mode. In general, economic growth rates of the low and middle income economies have been greater than those of the high income economies, with particularly rapid rates of growth in East Asia and the Pacific and in South Asia in general. The most spectacular example of changing economic fortunes in the world today is provided by China, home to one-fifth of the world's people. Changes in government policy in the late 1970s began a transformation of China's economy from a slumbering relic of central planning and socialism to a dynamic, rapidly growing blend of market-oriented capitalism and communism.

Historical and Projected Growth of World GDP
by Level of Economic Development
(annual percentage growth)

INCOME CATEGORY	1993–2000	2001–2008	2009–2012	2013–2016
High Income Economies	3.1	2.1	1.1	2.4
Low and Middle Income Economies	4.1	6.6	5.7	6.7
World	3.5	4.0	3.3	4.6

Source: IMF (2013)

The 1976 edition of the *World Book Encyclopedia* described China's economy in the following terms:

> *The Communist government controls and plans the economy of China. The nation's rulers are working to make China a highly industrialized nation. But they must overcome many problems to reach their goal . . . China's economy today cannot meet the basic needs of the huge, growing population for food, clothing, housing, and education. Factories of all types are needed. China has vast mineral resources, but most of them lie unused. Roads, railroads, and communications are still only partially developed. Some regions lack electricity and sanitation . . . The nation's economy still depends upon agriculture, and farming remains the leading industry. Agricultural products account for more than half the value of all the goods produced in China . . .*

A great deal has changed since those words were written. Rapid change began in 1978 when Deng Xioping came into power. Soon thereafter the Deng government began allowing farmers to sell their products in local markets, with the result that collective farms were effectively undermined. Agricultural output increased sharply. As growth continued, rural economies prospered, in part as a result of growth in farm income, but also because many farm families used their new-found income to invest in new enterprises. This set the stage for broader reforms across the economy, including the lifting of prohibitions against foreign investment and promotion of international trade.

During the period 1976–2015 China's Gross National Income (GNI) per capita increased on average by a phenomenal 9 percent per year, translating to a twenty-eight-fold increase in per capita GNI in a little more than a third of a century. Despite slightly lower rates of growth in the early years of the 21st century, the rate of economic expansion within China continued to be double or more that of other major world economies. This growth has led, in turn, to increases in personal income and living standards throughout China, and particularly in the large urban centers of the eastern and southern regions. The ability of Chinese citizens to consume has never been greater.

China's Gross National Income Per Capita, 1961–2015

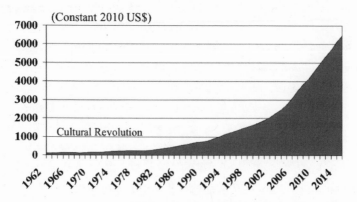

Source: World Bank Development Indicators Database (2017)

While China's successes have overshadowed those of their neighbors, a number of other countries have made notable economic gains over the past several decades.

- India, the world's second population billionaire, while continuing to be known as a poor country, has quietly spawned a large and modestly affluent middle class.
- Malaysia maintains much of its 19th century charm, with quiet villages, thatched-roof huts, and water buffalo in the fields, but also boasts world-class cities and shopping centers, manufacturing facilities, and transportation networks.
- South Korea has become an economic power in its own right. The same can be said of Taiwan.
- Thailand is rapidly becoming industrialized, with evidence of growing economic success visible throughout the countryside.

And so it goes. From Chile to the Philippines, and Estonia to Nigeria, the story is the same. Economic gains are being made in country after country, and with each passing day greater and greater numbers of people gain the ability to consume beyond the basic necessities. Large segments of the world economy are today characterized by rising per capita income, rising ability to consume, and rising per capita consumption of food and basic raw materials. The scale of change is unprecedented, the implications profound. But even as gains occur, per capita consumption in the economically emerging countries is far below that of the economically most developed. Using China, Egypt, India, the Congo (DRC), and Burundi as examples, average household consumption expenditure in these countries in 2013 was only 4.2, 3.6, 2.2, 0.7, and 0.4 percent of that of the United States.

Implications of Economic Growth

As the population increases around the world, an increasing percentage of that population has the ability to consume more on a per capita basis than in the relatively recent past. The combined effect of these two factors is a literal explosion in global consumption, and growing competition for the raw materials needed to support that consumption.

The relatively high rate of economic gain in developing nations, and in China and India in particular, is fundamentally changing the world economic landscape. In 1994, the economy of the United States (1) was 2.7 times larger than that of Japan (2), the world's second largest economy, over three times that of China's economy (3), and more than five times the size of India's economy (5). In the same year, Japan's economy was more than twice the size of India's.

World's 12 Largest Economies by GDP (PPP)—1994

Source: International Monetary Fund (2017).

Now fast forward just two decades to 2016. In the short span of just 22 years, the economy of China (3) had become the world's largest in terms of purchasing power, the size of India's economy (7) had eclipsed that of Japan (2), and the economies of both Brazil (9) and Indonesia (12) were larger than those of Italy, France, and the UK.

World's 12 Largest Economies by GDP (PPP)—2016

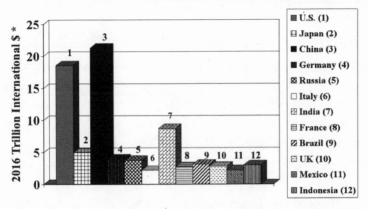

Source: International Monetary Fund (2017).

By 2050 many economists expect that China's economy will be the world's largest by a considerable margin and that the size of India's economy, though still smaller, will be within 10 percent of the U.S. It is further anticipated that Brazil's economy will eclipse that of Japan, Germany and Russia, becoming the world's 4th largest, and that the size of Mexico's and Indonesia's economies will each exceed that of any of the European countries. Moreover, the economies of Turkey and Nigeria are expected to be among the top fifteen. These changes have broad implications not only for global consumption patterns, but also for the ability of long-economically-dominant nations to compete for basic raw materials, including energy resources.

Perhaps anticipating changes in the global financial landscape, and seeking to ensure their own future participation in the global leadership hierarchy, the G-7 nations in 1999 invited a larger group of nations to participate in their discussions. Russia was formally added to the group in 1999, and more recently nations including Argentina, Australia, Brazil, China, India, Indonesia, South Korea, Mexico, Saudi Arabia, South Africa, and Turkey, along with a representative from the EU, have joined to create what is now known as the G-20.

Rising Raw Materials Consumption

Key Raw Materials

The population is growing rapidly, although at a decreasing rate. Consumption, in contrast, is growing more rapidly than population, and at an increasing rate. *One of the greatest environmental challenges for the decades ahead is how to accommodate rising consumption, and increasing competition for renewable and non-renewable natural resources, while also maintaining environmental quality.* Rising consumption translates to growing demand for raw materials of all kinds. Even though the efficiency of raw materials use is steadily improving, and recycling becoming more and more common, consumption of basic raw materials continues on an upward trend.

The principal raw materials used to produce structures, transport vehicles, and other durable and non-durable goods are relatively few in number. In fact, there are only four.

- portland and masonry cement
- wood
- metals (principally steel and aluminum)
- plastics

Some might also consider glass to be a key material. However, volumes used are only a tiny fraction of the four principal materials listed above. On a mass or weight basis, cement (and concrete of which cement is a part) is by far the dominant material used globally, with steel and wood used in lesser quantities.

World Consumption of Various Raw Materials, 2014

	BILLION METRIC TONS	BILLION CUBIC METERS
Cement*	4.180	1.25
Roundwood	1.656	3.68
Industrial Roundwood**	0.815	1.81
Steel	1.670	0.21
Plastics	0.311	0.34
Aluminum	0.054	0.01

* Cement is the active ingredient in concrete, binding sand and gravel. It makes up 8 to 15 percent of concrete by volume.

** The difference between roundwood and industrial roundwood is wood used for fuel. Roundwood includes both fuelwood and wood used in construction, and for making paper, furniture, and other wood products. Industrial roundwood does not include fuelwood.

Sources: Data for wood from FAO (2017); for cement from the European Cement Association (2016); for aluminum from the European Aluminum Association (2016);for steel from the World Steel Association (2016); and for Plastics from the Association of Plastics Manufacturers in Europe (2016).

In the U.S., consumption of concrete tops all other materials (remembering that cement accounts for only 8–15 percent of concrete). Wood is the next most used material, with quantities consumed annually greater than all metals and all plastics combined.

U.S. Consumption of Various Raw Materials, 2015

	MILLION METRIC TONS	MILLION CUBIC METERS
Roundwood*	189	420
Industrial Roundwood	185	411
Cement	93	30
Steel	110	15
Plastics	50.9	57
Aluminum	5.4	2.0

* Roundwood is the volume of all wood harvested.

Source: Data for wood from Forest Business Network (2016), with total roundwood estimated based on historical relationship; for cement, steel, and aluminum from the U.S. Geological Survey (2016); and for plastics from the American Chemistry Council (2016).

More, More, and Still More

When viewed from an historical perspective, society's use of minerals is a relatively recent development. The use of fossil fuels is almost breaking news when placed in perspective. Although mankind's discovery of the usefulness of metals goes back to about 330 BC and the Bronze Age, only in the last 200 years have substantial quantities of metals been consumed. According to Young (1992) "Large scale use of minerals began with the industrial revolution and grew rapidly for over two centuries. From 1750 to 1900 the world's use of minerals grew tenfold while the population doubled. Since 1900, it has jumped by at least thirteen-fold again."

The history of energy consumption is even more dramatic. For instance, world petroleum consumption in 2014 was about 227 times greater than in 1900 and more than four times greater than in 1960. Consumption of key materials used in construction of dwellings and other structures reveals a similar story. In 2014 the world population was 2.3 times greater than in 1960. Consumption of steel, aluminum, cement, plastic resins, and wood was 4.8, 11.9, 11.9, 51.8, and 1.8 times greater, respectively, than in 1960. In other words, consumption of all these materials, except wood, grew more rapidly—and in several instances much more rapidly—than population. Per capita consumption, therefore, increased.

World Growth in Consumption of Principal Raw Materials, 1960–2014

(POPULATION GROWTH DURING THIS PERIOD: 2.28X)

STEEL	CEMENT	ALUMINUM	PLASTICS	WOOD
4.75x	11.9x	11.9x	51.8x	1.8x

Source: Data for wood from FAO (2017); for cement from the European Cement Association (2016); for aluminum from the European Aluminum Association (2016); for steel from the World Steel Association (2016); and for plastics from the Association of Plastics Manufacturers in Europe (2016).

U.S. Flow of Raw Materials by Weight, 1900–2000

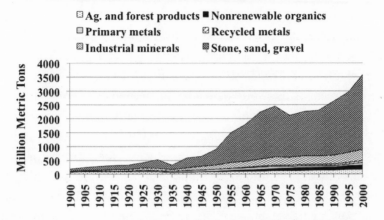

Sources: Wagner (2002); U.S. Geological Survey (2015a, b);
U.S. Geological Survey (2010); U.S. Census Bureau (1975).

The U.S. flow of raw materials illustrates the virtual explosion of raw materials use that began at the end of the Great Depression (1929–1939). Although slowed for brief periods in the course of economic corrections, the trend has been spectacularly upward.

In view of the rapid growth of global economies, it is perhaps fortunate that consumption per unit of economic activity (often referred to as intensity of use, or IU) tends to decrease after GDP reaches a certain level. Steel consumption in the United States, for instance, grew rapidly from the late 1800s through the early 1900s as the nation built miles upon miles of railroads, roadways, and bridges. However, as the national economy grew, and with it the development of basic infrastructure as well as increasing efficiency of steel use, a point was reached after which the IU of steel began to decline. Over the 85 year period 1888 through 1967, the IU of steel in the U.S. initially increased sharply, then leveled-off, and subsequently entered a prolonged trend of decline. Today IU values for a wide array of basic materials are rising in most of the developing nations, but are declining in the developed nations.

Changing Realities

High consumption rates have long characterized only a few nations. That this situation persists is evidenced by the high proportion of world raw material consumption on the part of leading economies. In 2010, for instance, residents of the most affluent nations, who directly accounted for 10.7 percent of the world's population, accounted for 59, 49, 45, 43, 42, 36, and 27 percent of all nickel, petroleum, aluminum, copper, lead, zinc, and steel, respectively, consumed worldwide. The same nations accounted for 53 percent of consumption of wood for industrial purposes and 67 percent of the world's consumption of paper. Similar disparity exists with respect to virtually all other materials. Moreover, consumption is even more lopsided than indicated by these numbers since considerable volumes of resources flow on a net basis to the most developed nations in the form of finished goods. Though long taken for granted by the developed nations, this disparity is unlikely to last for much longer.

Industrial Materials Consumption, Most Affluent vs. Other Nations, 2010.

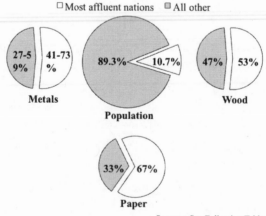

Sources: See Following Table.

That consumption patterns are changing can be seen by comparing data over a two-decade period, 1990–2010 (see Table on page 102). In just twenty years the percentage of world minerals consumption accounted for by the richest 12.1 percent (beginning of period) dropped dramatically. This is in part due to the shift of manufacturing from the U.S. and other developed nations to China and other developing nations, and in part due to increases in consumption within the developing countries themselves.

Part of the drop in world share of consumption between 2000 and 2010 can be attributed to the deep economic recession in the United States. However, shifts in consumption are not fundamentally occurring because of decreasing consumption in high income nations. They are, instead, occurring because of steeply rising consumption in a number of nations historically characterized by low consumption. Using China again as an example, increases in personal income have led to substantial growth in consumption of big ticket items such as automobiles and motorcycles, personal computers and color television sets, and water heaters and air conditioners. In 2014, for instance, 64 million air conditioners were sold in China, more than eight times as many as sold in the U.S. Consumption of countless other goods has risen sharply as well. While China is the poster child of such change, as noted previously, changing consumption patterns are by no means limited to that nation alone. Raw material consumption will continue to shift as nations now classified as economically developing become increasingly economically developed. It is a change of epic proportions.

China Consumption of Critical Materials 1990, 2000, and 2010 (percent of world consumption)

MATERIAL	1990	2000	2010
Steel	8.9	19.7	43.4
Aluminum	5.4	12.8	40.1
Copper	4.9	13.0	39.2
Nickel	4.2	6.6	29.3
Lead	7.4	9.0	44.1
Zinc	8.7	15.0	42.5
Cement	16.8	35.2	56.0
Industrial wood	7.1	14.7	16.4
Petroleum	3.6	6.3	11.1

Sources: Menzie-USGS (2012), Muriel-IADS (2013), U.S. Dept. of Energy-EIA (2015).

Population and Consumption of Basic Materials in Developed and Developing Countries as Percentage of World Total—1990, 2000, and 2010*

	% of world population			Steel			Aluminum			Copper		
							% of world consumption					
	1990	2000	2010	1990	2000	2010	1990	2000	2010	1990	2000	2010
High Income Countries*	12.1	11.3	10.7	46.8	51.3	26.6	68.4	63.9	44.8	66.9	58.1	42.6
Low Income Countries	87.9	88.7	89.3	53.2	48.7	73.4	31.6	36.1	55.2	33.1	41.9	57.4

	Lead			Zinc			Nickel			Industrial Wood		
				% of world consumption								
	1990	2000	2010	1990	2000	2010	1990	2000	2010	1990	2000	2010
High Income Countries*	69.6	63.3	41.8	60.9	55.3	36.4	70.9	69.1	58.8	66.0	62.1	52.6
Low Income Countries	30.4	36.7	58.2	39.1	44.7	63.6	29.1	29.9	41.2	34.0	37.9	47.4

	Cement			Petroleum		
	% of world consumption					
	1990	2000	2010	1990	2000	2010
High Income Countries*	34.6	25.3	11.3	56.8	51.4	49.0
Low Income Countries	65.4	74.7	88.7	43.2	48.6	51.0

* High income countries include the United States, Canada, 15 countries of northern and western Europe, Australia, New Zealand, Japan, and South Korea.

References

American Chemistry Council—Plastics. 2014. *U.S. Resin Production & Sales 2013 vs. 2012.*

Association of Plastics Manufacturers in Europe. 2015. *Plastics—the Facts 2014/15.*

Davis, L. 2015. "Air Conditioning and Global Energy Demand." *Berkeley Haas,* April 27.

European Aluminum Association. 2014. "Primary Aluminum Consumption 2011–2013." Accessed August 5, 2014.

European Cement Association. 2014. "Key Facts and Figures." Accessed August 5, 2014.

FAO. 2015. Forest Products Statistics. "Global Production and Trade of Forest Products—2013." *FAO Stat- Forestry Database.*

Frank, A. 1995. *Asian-based World Economy 1400–1800: A Horizontally Integrative Macrohistory.* University of Amsterdam. (November 12).

International Institute for Environment and Development (IIED). 2001. *Development of the Minerals Cycle and the Need for Minerals.* IIED Mining, Minerals and Sustainable Development Report No. 63. IIED/CRU International.

International Monetary Fund. 2015. "Data and Statistics." Accessed September 7, 2015.

Maddison, A. 2001. *The World Economy: A Millennial Perspective.* Organization for Economic Cooperation and Development.

Menzie, W. 2012. "China's Global Quest for Resources and Implications for the United States." *U.S. Geological Survey testimony before the U.S. China Economic and Security Review Commission,* Jan. 26.

Muriel, B. 2013. *China's Importance in International Commerce.* Institute for Advanced Development Studies.

PricewaterhouseCoopers. 2013. *The World in 2050: The BRICs and Beyond: Prospects, Challenges and Opportunities.* January.

Rogich, D. and G. Matos. 2008. *The Global Flows of Metals and Minerals.* U.S. Geological Survey. Open File Report 2008-1355.

United Nations. 2013. *World Population Prospects: The 2012 Revision.* Department of Economic and Social Affairs, Population Division.

U.S. Census Bureau. 1975. *Historical Statistics of the United States, Colonial Times to 1970—Part 2.*

U.S. Department of Energy. 2015. "International Energy Statistics—Petroleum." Energy Information Administration, Accessed September 7, 2015.

U.S. Geological Survey. 2015a. Commodity Statistics and Information, Accessed October 10, 2015.

U.S. Geological Survey. 2015b. Historical Statistics. [Available for all commodities] Accessed October 10, 2015.

Wagner, L. 2002. *Materials in the Economy: Material Flows, Scarcity and the Environment.* U.S. Geological Survey Circular 1221.

World Bank. 2015. "Development Indicators Database," Accessed August 1, 2015.

World Bank. 2016. "Household Final Consumption Expenditure per Capita Database."

World Steel Association. 2015. "World Steel in Figures 2014." Accessed September 7, 2015.

Young, J. 1992. "Mining the Earth." *Worldwatch Paper 109* (July). Washington D.C.: Worldwatch Institute.

Zhao, X., N. Li, and C. Ma. 2011. *Residential Energy Consumption in Urban China.* University of Western Australia, School of Agricultural and Resource Economics, Working Paper No. 1124.

PART 3

RESOURCE REALITIES

Economic prosperity, social well-being, and environmental sustainability ultimately depend upon resource availability—food, potable water, energy, and basic materials from which physical goods are made. The three chapters that follow focus on physical resources—mineral resources, biological resources, and those used in producing energy.

CHAPTER 7

Minerals and Metals Basics

*We are running out not of mineral resources but of ways
to avoid ill effects of high rates of exploitation.*
—DAVID BROOKS AND PETER ANDREWS

When you pick up your knife and fork in preparation for attacking a T-bone steak you probably don't think about the source of metals used in making the utensils, the quantity of energy expended, or the ingenuity and sweat of the miners who extracted the ore. It is similarly unlikely that you think about what went into creating the plate, the steak, the table they sit on, or the floor beneath. We tend to take these things for granted. Yet, each of us consumes vast quantities of wood, metals, non-metallic minerals, plastics, glass, and other materials each year.

This chapter deals specifically with our use of minerals and metals. Non-metallic minerals include limestone, from which cement is made, sand, salt, clay, marble, and phosphorous, the principal use of which is as an ingredient in fertilizer. Included among the metallic minerals are familiar resources such as aluminum, steel, copper, titanium, and tungsten, and more exotic materials such as lutetium, used principally in LED light bulbs and integrated circuits, and erbium, a critical element in amplifiers in fiber-optic data transmission.

Consider that each person in the U.S. consumes, on average, 38,524 pounds of minerals every year—a figure that includes newly mined minerals, metals, and fossil fuels. That is over 19 tons per person. When recycled metals are included, the total is even higher. Assuming that per capita consumption of minerals remains the same as today, each baby born in the U.S. will consume about 3.1 million pounds of newly mined minerals, metals, and fuels over the course of a lifetime. This is higher than any other country and far higher than the world average.

As noted in the preceding chapter, high consumption of minerals and other raw materials by the U.S. and other developed nations has long been viewed as the natural order of things. Continued availability of resources has been assumed as well, even as population and the global economy have continued to expand. Now, however, rising global competition for minerals, net import dependence of the U.S. and other developed nations, and environmental implications of rapidly increasing minerals extraction are triggering concerns about the future.

Estimated Average Minerals Consumption over the Lifetime of Every American Born Today

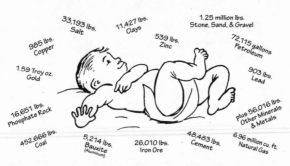

3.11 million pounds of minerals, metals, and fuels in their lifetime

Courtesy of Minerals Education Coalition (2016)

Minerals Abundance

Some of the most commonly used metals, such as iron, silicon, aluminum, magnesium, titanium, and manganese are found in great abundance within the Earth's crust. Silicon (sand), for example, makes up more than a quarter of the crust, with aluminum (in the form of alumina and bauxite) and iron accounting for about another eight and five percent. The top twelve elements account for over 99 percent of the mass of the crust.

Other minerals are relatively scarce, collectively comprising less than

1 percent of the crust's mass. For instance, gold, silver, copper, tin, lead, and zinc are also commonly-used metals, but they are far less abundant than silica, aluminum and iron. In fact, each accounts for less than one one-hundredth of one percent of the mass of the Earth's crust.

Relative Percentages by Weight of Chemical Elements in the Earth's Crust

ELEMENT	PERCENT MASS
Oxygen (O)	49.2
Silicon (Si)	25.7
Aluminum (Al)	7.5
Iron (Fe)	4.7
Calcium (Ca)	3.4
Sodium (Na)	2.6
Potassium (K)	2.4
Magnesium (Mg)	1.9
Hydrogen (H)	0.9
Titanium (Ti)	0.6
Chlorine (Cl)	0.2
Phosphorous (P)	0.1
Manganese (Mn)	0.1
All Others	0.7

Source: Rogers, et al. (2000).

One group of metals is both critically important and extremely rare. Appropriately named *rare earth metals,* these provide essential components in a wide range of modern electronic technologies, ranging from TVs, fluorescent light bulbs, cell phones and computers to "green" magnets in electric motors that power hybrid cars and generators used in wind turbines. Rare earths are also essential to medical diagnostic equipment and almost all military systems.

The matter of abundance is somewhat relative. One way to look at this is by considering occurrence of minerals in the context of the mass of the crust of the Earth, estimated at approximately 62 quintillion (62 followed by 18 zeros) tons. Aluminum and iron are obviously abundant.

But even one one-hundredth of one percent of 62 quintillion tons, the estimated occurrence of copper, is a fairly large number—6.2 trillion tons. So, even though copper is considered to be relatively scarce, it could also be said to be abundant. Based largely on the vastness of potential supply, conventional wisdom holds that most minerals, though non-renewable, will never run out in the physical sense. However, it is worth noting that for precisely the same reason, forests were viewed as inexhaustible up to about 125 years ago. Moreover, much of minerals volume is located far below the Earth's surface and/or below the oceans.

A measure of the reserves of a raw material is the "reserves life index." This number, expressed in years, indicates known reserves that are economically accessible using current technology, divided by current annual consumption. The index does not consider possible increases in future consumption, nor is consideration given to technology improvements or potential discoveries of additional ore deposits. The reserves life index thus provides a snapshot of the reserves status at any point in time.

Reserves Life and Reserves Base Life Indices for Common Metals, 2005

METAL	RESERVES LIFE INDEX (YEARS)	RESERVE BASE LIFE (YEARS)
Aluminum (Bauxite)	>100	>100
Platinum & Palladium	>100	>100
Iron	52	>100
Molybdenum	43	94
Nickel	40	93
Copper	29	57
Zinc	20	42
Lead	19	39
Tin	18	33
Gold	15	33
Silver	12	26

Source: Hewitt (2006).

Reserves life index values are sometimes misinterpreted to mean that the world is on the brink of running out of various raw materials. A life index value for tin of 18 years is, for example, not reassuring to anyone concerned about resource depletion. However, in reality, the indices tend to remain relatively constant or even increase over time as new technologies and new ore discoveries, and/or changing economic variables expand the supply of economically available materials.

The reserves base life numbers in the table above take into account measured deposits as well as those indicated by early exploration, and include economically proven and probable portions of these deposits as well as marginal and sub-economic reserves. As economically available reserves of any given mineral dwindle, prices rise, making previously sub-economic reserves feasible. In addition, rising prices provide market incentives for further exploration and technology development that can bring further resources into play. Rising prices can also encourage development of new technologies that allow substitution, such as fiber optic cable for copper wire.

Reserves indices values may decline as ore is mined and if reserves in at least replacement quantity are not found. More commonly, however, values change little or increase even in the face of substantial extraction. For example, in 1970, world copper resources reserves were estimated to be 280 million metric tons. Over the following 42 years, more than 400 million metric tons of copper were produced worldwide, but world copper reserves in 2012 were estimated to be 680 million metric tons, more than double those in 1970, despite the depletion by mining of considerably more than the original estimated reserves. Similarly, predictions of "peak oil" were substantially modified when exploitation of tar sands became economically viable.

Ore Quality

Just as reserves life numbers are often misunderstood, declining ore quality is commonly, and erroneously, interpreted as indicating lower quality of metals produced. But "ore quality" simply refers to the concentration of metals within a given quantity of ore. Use of lower quality ores does not translate to lower quality metals. A copper pot made from ore mined today is every bit as good as a copper pot made 50 years ago. Ore quality is determined solely based on how much metal is contained within a given quantity of ore.

The term "ore" applies to rock that contains a sufficient concentration of minerals (usually metallic minerals) to make extraction economically attractive. Most mining for metallic ore is done in open-pit mines, meaning that a hole or pit is dug to get at a seam of ore. The ore typically lies beneath the Earth, and so mining involves not only extraction of ore, but removal of earth and associated vegetation that sits above. The result is that extraction of a ton of ore can require the movement of 20 to 30 tons or more of earth. The ore itself occurs in varying concentrations, meaning that to obtain a ton of metal requires removal of even more material. The highest concentrations, in which metals make up a high percentage of the weight of an ore, are referred to as high quality ores; lower quality ores contain a lesser percentage of metal.

The magnitude of ore extraction varies widely depending upon what is being mined and the ore concentration. For instance, consider iron, gold, and diamonds.

- Iron. As is widely known, iron is used to make steel. Recall that iron makes up 4.7 percent of the Earth's crust. However, the percentage of iron in ore ranges from as high as 70 percent in some geographic areas to as low as 28 to 33 percent in others. So, to get one ton of iron requires extraction of 1.4 to 3.3 tons of ore. However, this is not the only raw material used in making steel. A number of other metals—including chromium, nickel, zinc, molybdenum, manganese, and vanadium—are used in making various alloys of steel that enhance strength, hardness, and other properties. Zinc is also used as a coating in galvanized steel. The ores of these metals are far less concentrated than iron, magnifying environmental impacts of extraction and processing.

- Gold. In contrast to iron mining, profitable mining of gold, a metal that comprises only 0.0000002 percent of the Earth's crust, requires that the gold must be 4,000 to 5,000 times more concentrated in a seam of ore than in the Earth's crust in general. Even then, a ton of such an ore would contain only one-third ounce of gold.

- Diamonds. The story of jewelry-grade diamonds is even more dramatic. In this case, it is necessary to process 250 *tons* of ore to obtain one carat of jewelry-grade diamond! It is easy to overlook the enormous environmental impact that underlies the beautiful radiance of a pair of diamond earrings.

The quantity of high quality ore available worldwide is relatively low compared to the quantity of lower quality ore. In general, the most easily accessed and highest quality ores are the most economical to extract and process, and are thus the most likely to be mined first. As the highest quality ores are depleted, attention shifts to ores of progressively lower concentration (i.e., to lower and lower ore quality). Such shifts result in impacts over progressively larger land areas and in greater expenditure of energy per unit of ore obtained. An example of the impact of declining ore quality is provided by copper. Since the late 1800s the concentration of copper in prime mining areas has declined from 2 percent to about 0.6 percent, meaning that it is necessary today to mine over three times as much ore as in 1880 to obtain the same quantity of copper. Similar trends in deteriorating ore quality can be seen for nickel, gold, and platinum group metals.

Relative Availability of Metallic Ore of High and Low Quality

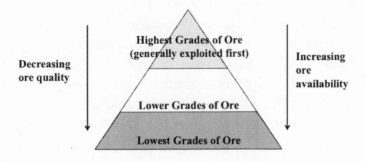

Decreasing ore quality

Highest Grades of Ore (generally exploited first)

Increasing ore availability

Lower Grades of Ore

Lowest Grades of Ore

Minerals Availability

Limits Imposed by Physical Supplies

Discussions of future mineral resource supply are largely defined by two diametrically opposed views:

- Because of rapidly rising consumption, and the fact that minerals are both finite in supply and non-renewable, the world will soon run out of mineral resources.
- Since economically accessible stocks of minerals tend to remain steady or increase rather than decrease over time as technologies for accessing and processing lower grades of ore are developed, there is little likelihood of exhausting the world's physical mineral resources within the foreseeable future or perhaps ever.

The former view was advanced in the early 1970s book *Limits to Growth* and more recently by a sequel to that work. Near-term non-renewable resource exhaustion is also forecast in *Scarcity: Humanity's Final Chapter?* in which non-renewable resource scarcity is described as "the most daunting challenge to ever confront humanity."

Others see things differently. For instance, scientists with the British Geological Survey concluded in 2012 that ". . . physical exhaustion of primary metal resources is unlikely . . . With scientific understanding improving over the next 40 years, reserves will replenish from previously undiscovered resources." Similarly, scientists at the University of Tennessee and Oak Ridge National Laboratory wrote in 2014 that "Mineral resources are sufficient for our current technology needs for the next century." These conclusions were qualified with observations that such factors as unforeseen increases in minerals consumption, insufficient investment in technological development, increased energy prices, and environmental and political constraints could all adversely affect future minerals supply. A need for continuing advances in science and technology and greater implementation was also emphasized in these and other assessments.

John Tilton, a distinguished professor of mineral economics at the Colorado School of Mines, succinctly explained the prevailing view among minerals experts:

> *We now know that the world is not likely to wake up one day to find the [minerals] cupboard bare or the well dry. We will not run out of mineral commodities the way a car runs out of gasoline: one minute speeding along the highway, the next completely stranded on the berm. Depletion, if it becomes a serious problem, will raise the real costs of finding and producing mineral commodities, but probably slowly yet persistently over years and decades. There likely will be signs of impending scarcity long before there actually are serious shortages.*

Tilton, however, cautioned that:

> *. . . the past is not necessarily a good guide to the future . . . , and that it simply isn't known . . . whether coming generations face a future of mineral commodity shortages. Those who argue otherwise ask the rest of us of share their faith, or lack of faith, in technology.*

The bottom line here is that there is a difference between the *physical existence* of a given mineral resource and *availability* of that resource. While physical exhaustion of resources is unlikely, availability sufficient to supply growing societal demands is nonetheless not assured. In this regard, there are serious concerns.

A select United Nations panel reported in 2011 that, driven by rapidly increasing metals consumption in Asia, global minerals extraction rose from 3.9 to over 9 billion tons between 1980 and 2008. As much as a threefold increase in materials consumption was projected by 2050 as compared to 2000. A more recent assessment by the European Commission indicated the likelihood of "massive growth in global demand for mineral resources as the human population progresses towards the 9 billion mark . . . and as populations in developing countries aspire to enjoy infrastructure, services and goods currently enjoyed by wealthier nations." This assessment further concluded that "the quantity of minerals to be produced between now and 2050 is expected to greatly exceed the quantities extracted between the onset of humanity and now."

Putting specific numbers on these projections, Rio Tinto, the global mining giant, indicated in 2010 that global demand for the base metals copper, aluminum, and iron would increase by a factor of 2.5, 3.3, and 3.8 times, respectively, over the thirty-year period 2000–2030. A European assessment projected increases of similar magnitude for other critical minerals over the same time period, projecting annual increases in global demand of more than 4 percent per year for 14 critical metals, a growth rate translating to doubling times of less than 18 years. All of this is raising doubts about whether future availability of mineral resources will be sufficient to meet societal demands.

Concerns about rapidly rising consumption and increased competition for minerals are especially great among the most highly economically developed nations—specifically the United States and western European nations, which are historically among the highest per capita consuming nations as well as the largest scale net importers of raw materials. The potential for scarcity of mineral resources is increasingly perceived as a risk by both governments and the private sector. Concerns are also reflected in various reports and articles:

"Critical minerals: Growing demands, rising tensions" (Center for a New American Security, January 3, 2011)
"Minerals and metals scarcity in manufacturing: The ticking time bomb" (PricewaterhouseCoopers, December 2011)
"Dangerous dependence: US increasingly beholden to imported raw material" (*Detroit News,* April 5, 2012)
"Rare Mineral Shortage Could Upset Global Markets" (*Industry Week,* November 13, 2015)

In a 2011 global survey of 69 leading companies, resource scarcity was identified as a major concern in every industry represented. In 2014 the World Economic Forum indicated that "scarcity of mineral resources could endanger future consumption as demand continues to climb."

As this situation unfolds, attention within the developed nations is focusing on where known resources are concentrated. Evaluation has determined that:

- A large portion of available mineral deposits globally are located in the developing countries.
- Well over half of global exploration for minerals is accordingly going to the developing countries.
- Developing countries have accounted for almost all increases in global mineral production in recent years.

An assessment of the European Union's Polinares Project concluded that, as a consequence of these trends, "the developed countries of the world, traditionally the dominant consumers of minerals, as well as major producers, are finding themselves progressively marginalized."

Interestingly, angst over rising competition for minerals was evident over 60 years ago among top U.S. governmental officials, although at that time the focus was simply on minerals consumption by a select group of developed nations. The topic of minerals and energy availability was foremost in thinking, when in 1951 President Harry Truman created the President's Materials Policy Commission, charged with studying the materials situation of the United States and of the Free World. The final report referred to a "large and pervasive materials problem." Concerns were again evident almost 20 years later (1970) as Congress created, with President Richard Nixon's blessing, a National Commission on Materials Policy. Made up of representatives of the private sector and government, the new commission was charged with developing a national raw materials policy and recommendations for implementing such a policy.

Included in the introductory remarks of the final report of the Commission is the following statement:

Now with established economic strength and foreign exchange reserves, they [the other industrialized nations] are searching actively for stable supplies of minerals and fuels. Some US enterprises that have long pursued the same materials have reacted with alarm to this competition. This business competition may evolve into a mutually destructive race for resources when combined with rapidly growing demand for materials.

The document also raised questions about rising U.S. import dependence for basic materials, and the ability of the U.S. to sustain related payments. Reemphasizing growing foreign competition for energy and mineral resources, report authors indicated that domestic resources needed to "play their proper role in the global supply-demand balance," called orderly development of domestic resources a high priority, and proposed efforts to accelerate waste recycling and increase efficiency of raw materials use.

Despite the findings and advice of the Commission, domestic production of minerals declined significantly and net U.S. import dependence for minerals rose dramatically in the 45 years following 1970. Modest improvements were realized in the rate of metals recycling. Even in the face of these developments, as discussed extensively in the book *Mineral Economics and Policy,* mineral shortages did not subsequently develop, nor did minerals prices rise, as might be expected in the face of declining availability. In fact, the quantity of known reserves rose for most minerals even as consumption increased, leading to reductions rather than increases in price. These outcomes were surprising to many.

Limits Imposed by Environmental and Social Concerns

In the early 1970s, as part of an assessment of relative minerals abundance and future availability, several Canadian mining experts commented:

We argue that it is naive to jump from the premise that mineral resources are physically finite to the conclusion that this limits their economic availability ... We further argue that it is equally naive to jump from the premise that mineral resources are economically infinite to the conclusion that they could be produced in enormous volume without major social or political problems.

They concluded with the observation that:

The tendency for environmental impacts to rise as ore quality and accessibility decline suggests that there are limits to metals consumption even if physical availability is not limited.

Environmental impacts are realized in every phase of mining: exploration, building of access roads, removal of overburden, extraction, beneficiation (separation and disposal of waste material), metallurgical processing and refining, and mine decommissioning. Impacts on and near mining sites are often large, scarring the landscape, releasing particulates, and polluting waterways—sometimes for decades following cessation of mining activity. Less obvious are impacts linked to high energy consumption throughout the process of extraction and conversion to products.

For a long time environmental damage and adverse social impacts associated with mining were accepted as simply a fact of life. Yes mining was dirty, often dangerous and unhealthy, and likely to leave an enduring legacy, but metals were essential to society, mining provided jobs and income, and that was that. Over time, however, as people became more informed and aware of the nature of damage, attitudes toward mining began to change. In the words of several Canadian writers, society has come to the collective recognition that it is not acceptable to cause irreversible damage, and that many past activities were undertaken legitimately and in good faith, then later found to have created long-lasting harmful effects. The result was increasing citizen resistance to mining activity that was revealed in ever greater regulatory oversight and permitting delays and restrictions, especially within the developed countries. One effect was to encourage outsourcing of minerals to locations with fewer constraints to operation.

More recently, concerns about environmental impacts of mining have gone global, triggered by the explosion of minerals consumption over the past several decades. Many have also voiced alarm about future availability of minerals in view of the ongoing, dual and reinforcing effects of continuing population growth and rapidly rising consumption. Consequently, a major effort to drastically reduce virgin minerals consumption has been initiated. Led by the United Nations Environment Program and the Organization for Economic Cooperation and Development, the idea is to decouple consumption and resource use by greatly increasing the efficiency of use of physical materials such as critical minerals. A central premise is that nations will cooperate to find ways to reduce impacts and that the high-consuming nations, in particular, will do whatever they can to extend the use of minerals at their disposal. Improvements in the efficiency of raw materials conversion and use, and greater reliance on metals recovery and recycling, are among the strategies proposed.

Minerals Import Dependence

As global minerals consumption undergoes rapid change, the developed nations find themselves heavily reliant on the developing nations for mineral supplies. The world's major consuming nations, the United States, Japan, EU nations (and particularly western European nations), and now China are all major net importers of metallic and non-metallic minerals. The United States provides a case in point. Early in the 21st century the U.S. appears to be in a vulnerable position as a net importer of not only minerals and metals, but all categories of raw materials used to support the American economy and lifestyle. On the net import list are most metals, Portland and masonry cement, petroleum (the basis for most plastics), and wood. In many cases, the level of net imports is massive.

Net U.S. Imports of Selected Materials as a Percentof Apparent Consumption—2015, and by Major Foreign Sources a,b,c,d,e

MATERIAL	% IMPORTED	PRINCIPAL FOREIGN SOURCES (2011–2014)
Niobium (Columbium)	100	Brazil, Canada
Manganese	100	South Africa, Gabon, Australia, Georgia
Graphite	100	China, Mexico, Canada, Brazil
Strontium (Celestite)	100	Mexico, Germany, China
Arsenic	100	Morocco, China, Belgium
Bauxite	100	Jamaica, Guinea, Brazil, Guyana
Fluorspar	100	Mexico, China, S. Africa, Mongolia
Indium	100	China, Canada, Belgium, S. Korea
Thallium	100	Germany, Russia
Thorium	100	India, France
Asbestos	100	Brazil, Canada
Quartz crystal (industrial)	100	China, Japan, Romania, UK
Rubidium	100	Canada
Cesium	100	Canada
Tantalum	100	China, Germany, Indonesia, Kazakhstan
Mica sheet (natural)	100	India, Brazil, China, Belgium
Scandium	100	China
Vanadium	100	Czech Rep., Canada, S. Korea, Austria
Iodine	100	Chile, Japan
Gallium	100	Germany, China, UK, Ukraine
Gemstones	99	Israel, India, Belgium, S. Africa
Bismuth	95	China, Belgium, Peru, UK
Titanium mineral concentrates	91	South Africa, Australia, Canada, Mozambique
Platinum	90	South Africa, Germany, UK, Canada
Garnet (industrial)	88	Australia, India, China
Germanium	85	China, Belgium, Russia, Canada
Antimony	84	China, Bolivia, Belgium, Thailand
Diamond (dust, grit, powder)	84	China, Ireland, Romania, S. Korea
Potash	84	Canada, Russia, Israel, Chile
Stone (dimension)	83	China, Brazil, Italy, Turkey

MATERIAL	% IMPORTED	PRINCIPAL FOREIGN SOURCES (2011–2014)
Zinc	82	Canada, Mexico, Peru, Australia
Lithium	80+	Argentina, Chile, China
Rhenium	79	Chile, Poland, Germany
Silicon carbide (crude)	77	China, S. Africa, Netherlands, Romania
Rare earth metals	76	China, Estonia, France, Japan
Cobalt	75	China, Norway, Finland, Russia
Tin	75	Peru, Indonesia, Bolivia, Malaysia
Silver	72	Mexico, Canada, Poland, Peru
Barium (Barite)	70	China, India, Morocco, Mexico
Peat	69	Canada
Titanium (sponge)	68	Japan, Kazakhstan, China
Chromium	66	South Africa, Kazakhstan, Russia, Mexico
Palladium	58	Russia, S. Africa, UK, Switzerland
Tungsten	49	China, Bolivia, Canada, Germany
Magnesium compounds	43	China, Brazil, Canada, Australia
Aluminum	40	Canada, Russia, United Arab Emirates
Mica, scrap/flake-natural	39	Canada, China, Finland, Mexico
Silicon	38	Russia, Brazil, China, Canada
Nickel	37	Canada, Australia, Russia, Norway
Copper	36	Chile, Canada, Mexico
Salt	32	Chile, Canada, Mexico, The Bahamas
Lead	31	Canada, Mexico, Peru, Australia, Kazakhstan
Nitrogen (fixed), Ammonia	29	Trinidad/Tobago, Canada, Russia, Ukraine
Petroleum (crude & refined)	27	Canada, Saudi Arabia, Mexico, Venezuela, Iraq
Magnesium metal	26	Israel, Canada, China, Mexico
Iron and Steel	25	Canada, S. Korea, Brazil, Russia
Softwood Lumber	25	Canada
Perlite	21	Greece, Turkey
Pumice	21	Greece, Iceland, Mexico
Vermiculite	20	S. Africa, Brazil, China
Sulfur	16	Canada, Mexico, Venezuela
Gypsum	14	Mexico, Canada, Spain

MATERIAL	% IMPORTED	PRINCIPAL FOREIGN SOURCES (2011–2014)
Talc	13	Pakistan, Canada, China, Japan
Feldspar	12	Turkey, Mexico, Germany, India
Beryllium	11	Kazakhstan, China, Nigeria, UK
Iron and steel slag	11	Canada, Japan, Spain, Italy
Portland and masonry cement	10	Canada, S. Korea, China, Greece

a Principal foreign sources arranged by most important supplier to the left, next most important supplier to the right of that, and so on.
b Minerals data from U.S. Geological Survey. 2016. Mineral Commodity Summaries—2016.
c Petroleum data from U.S. Department of Energy, Energy Information Administration, 2016 (Feb.).
d Softwood lumber data from RISI, 2015.
e Also on the net import list are uranium, industrial diamonds, lime, phosphate rock, construction sand and gravel, and crushed stone. The net export list is relatively short, including metallic abrasives, alumina, boron, cadmium, clays, diatomite, gold, helium, iron and steel scrap, iron ore, kyanite, molybdenum, industrial sand and gravel, selenium, soda ash, titanium dioxide pigment, wollastonite, and zeolites.

Many view the current level of U.S. minerals import dependence as both alarming and unnecessary. They point out that regulatory compliance requirements and policy uncertainty are among a number of factors that influence where extractive activity occurs in the world. Mining interests similarly point to lengthy delays in permitting processes within the U.S. as a key factor in limiting domestic mining activity. The mining consulting company Behre Dolbear ranks the U.S. highly as a place for mining investment, but also commented in 2015 that:

The U.S. is a bit of a paradox. Its experienced governmental and financial sectors contribute to high investor confidence. However, the public's worry over environmental legacy, combined with the incremental concern that new production capacity represents, leads to an onerous permitting process that creates sufficient uncertainty to sometimes destroy the viability of new projects.

Numerous industry advocates believe that changes in public policy vis-à-vis permitting and regulations are needed. Others disagree, saying that strong regulatory oversight provides competitive advantage rather

than the other way around, that richer deposits and more economic mining possibilities are the factors responsible for shifting mining outside
U.S. borders, and that any weakening of current law would be bad for
both the industry and the environment.

These strongly divergent views, along with negative public attitudes
toward mining, largely define the politics of resource extraction within
the United States. Meanwhile, minerals flow into the United States on
a daily basis and on a massive scale, with environmental impacts of extraction having occurred outside U.S. borders. The same situation exists
in the EU.

The Role of Recycling in the Minerals Picture

No discussion of minerals resources would be complete without consideration of recycling. Recycling not only reduces the need for mining and
processing of virgin ore, but also reduces considerably the consumption
of energy in metal products manufacturing. The Steel Recycling Institute
reports that every ton of steel recycled conserves 2,500 pounds of iron
ore, 1,400 pounds of coal, and 120 pounds of limestone. An estimated
74 million tons of scrap iron and steel were recycled in the U.S. in 2015.
The environmental benefits are enormous. As indicated at the beginning
of this chapter, an average American consumes over 19 tons of virgin
mineral resources each year. About 14 of these tons are non-fuel minerals.
Total consumption is even greater when recycled content is considered.
How to shift more of our consumption to recovered, rather than virgin,
materials is a question that many are seeking answers to.

Seeking Reductions in Virgin Materials Extraction

A 2011 UNEP study of global recycling rates for 60 metals and metalloids
(elements that have properties of both metals and non-metallic minerals)
revealed that while end-of-life recycling rates of the most commonly used
metals (including iron, copper, nickel, zinc, lead, manganese, cobalt, and
gold) are above 50 percent, recycling rates for most other metals are far

below that. Of the 60 metals examined, recycling rates for over two-thirds were less than 50 percent, and for over one-half of them, less than 1 percent. Most of the metals and metalloids that are seldom, if ever, recovered for recycling tend to be those used in small amounts for very precise technological purposes such as high-strength magnets and computer chips. They are also typically elements found in very low concentrations in nature, translating to high environmental impacts in extraction and processing. At present, technical and economic issues are the primary barriers to recovery.

The finding that so many metals are not being recovered for recycling, and at a time when global competition for minerals is rapidly rising, is the basis for the UNEP/OECD initiative to reduce resource consumption, especially within high consuming nations, and to achieve closed-loop recycling of resources. Dubbed "decoupling" or, in some circles "dematerialization," the idea is to uncouple rising consumption from needs for more and more virgin resources, relying instead on resource recovery and reuse and technological advances to increase efficiency of materials use. The end goal is a drastic reduction in extraction of virgin raw materials within the relatively near term.

A goal of increasing recovery and recycling rates, and of substantially increasing efficiencies of materials use, is straightforward. Actually achieving this goal, however, will be a significant undertaking.

Improving Materials Use Efficiency

Improving materials use efficiency is not a new idea, and in fact, is a goal pursued on a daily basis by industries all over the world. Competitive pressures alone act as a constant incentive for business and industry to provide goods and services at lower cost, often translating to improvements in materials use efficiency and reduction of waste.

It is competition-driven incentives that have led to increasingly lighter and less resource intensive computers and other electronic products, vast reductions in the thickness of steel and aluminum beverage cans and plastic bottles, reduced basis weights of paper products, development of lightweight structural concrete, orders of magnitude improvements in

voice and data transmission capacity from previous use of copper wire
to use of thinner and lighter optical fiber, and so on. While all of this is
quite impressive, sharp growth in metals consumption is occurring despite these gains. Consequently, the improvements that decoupling advocates are calling for will require progress well beyond what is being
accomplished already.

Increasing Recovery and Recycling

Increasing waste recovery and recycling of materials recovered is another significant challenge. Steel recycling provides a case in point. Steel
is both the most-used metal in the world, as well as the most recycled.
Yet, there are a number of problems associated with steel recycling that
remain unsolved. A significant problem in recycling of all metals is
contamination resulting from mixing with other metals and other materials. With regard to steel, the major source of contamination of steel
discards is the steelmaking process itself. There are as many as 3,500
different grades of steel made by adding various metals to steel in the
form of alloying elements. The major problem with contaminants is that
some of them are very difficult to remove regardless of the recycling
process used, resulting in undesirable properties in the steel produced.
Moreover, each time scrap steel is re-circulated the concentration of re-
siduals rise, thereby making processing more difficult and potential uses
more restrictive.

Recovery of discards for recycling presents another challenge. Using
the U.S. experience as an example, of all ferrous material discarded
during the period 2004–2009, only 54 percent was recovered for recycling. About 30 percent of discards were deemed unrecoverable.

So while metals recovery plays an important role in the metals picture,
systems today are far from perfect, and visions of closed-loop infinite use
recycling and reuse remain elusive. Even without growing consumption,
mining of virgin metals remains an important factor in providing the metals needed by society.

Considering the Options

Until recently, the developed nations have faced little competition for the world's mineral resources, while at the same time experiencing little of the adverse environmental and social impacts of minerals extraction and processing. Rapid economic and population growth in Asia and elsewhere has radically changed the world resources equation.

In light of the anticipated increase in environmental and social problems linked to unrestrained and rapidly increasing consumption of minerals, a relevant question today is what to do about it. One proposed solution is drastic reduction in virgin raw materials extraction through greater attention to recycling and improvements in materials use efficiency—and especially in high-consuming countries. Another is action to reduce consumption through such things as significantly extending the life of personal computers and electronics, reducing packaging, and developing options to reliance on personal transportation. Others have proposed policy changes to allow greater mining activity within those countries responsible for the bulk of minerals consumption. In the meantime, technology limitations dictate high inputs of virgin raw materials even when recycling rates are relatively high.

References

Behre Dolbear Group, Inc. 2015. "Where to Invest in Mining—2015."

Bloodsworth, A., and G. Gunn. 2012. *"The Future of the Global Minerals and Metals Sector: Issues and Challenges out to 2050." Geosciences: the BRGM Journal for a Sustainable Earth,* June, 90–97.

Brooks, D., and P. Andres. 1973. "Mineral resources, economic growth, and world population." In *Materials—Renewable and NonRenewable Resources,* edited by H. Abelson and A. Hammond, 41–47. American Association for the Advancement of Science, Special Science Compendia No.4.

Clugston, C. 2012. *Scarcity: Humanity's Final Chapter?* Booklocker.com, Inc.

Damath, R. 2010. *Iron and Steel Scrap: Accumulation and Availability as of December 21, 2009.* Institute of Scrap Recycling Industries, Nov. 23.

Earthworks. 2012. "Attractiveness of the United States for Mineral Investment," Accessed May 15, 2015.

Euractiv. 2010. "Raw Materials Headed for a Global Resource Crunch?" January, Accessed May 15, 2015.

European Commission. 2014. *Report on Critical Raw Materials for the EU.* Report of the Ad-hoc Working Group on Defining Critical Raw Materials.

European Environment Agency. 2011. *Increasing Global Competition for Decreasing Stocks of Resources.*

Fischer-Kowalski, M., M. Swilling, E. von Weizsäcker, Y. Ren, Y. Moriguchi, W. Crane, F. Krausmann, N. Eisenmenger, S. Giljum, P. Hennicke, P. Romero Lankao, A. Siriban Manalang, and S. Sewerin. 2011. *Decoupling Natural Resource Use and Environmental Impacts from Economic Growth- A Report of the Working Group on Decoupling to the International Resource Panel.* United Nations Environment Program (UNEP).

Graedel, T., J. Allwood, J-P Birat, B. Reck, S. Sibley, G. Sonnemann, M. Buchert, and C. Hagelüken. 2011. *Recycling Rates of Metals—a Status Report.* Working Group on Global Metal Flows to the United Nations International Resource Panel, United Nations Environment Program (UNEP).

Hewitt, M. 2006. "Global Metal Reserves." DollarDaze, Accessed November 3, 2014.

Humphreys, D. 2012. *Mining Investment Trends and Implications for Minerals and Availability.* European Commission, Polinares Project, working paper 15, March.

Institute of Scrap Recycling Industries. 2012. *The ISRI Scrap Yearbook 2012.*

Kesler, S. 2007. "Mineral Supply and Demand into the 21st Century." In *Proceedings, Workshop on Deposit Modeling, Mineral Resources Assessment, and Sustainable Development.* U.S. Geological Survey.

Meadows, D. H., D.L. Meadows, J. Randers, and W. Behrens III. 1972. *The Limits to Growth: a Report for the Club of Rome's Project on the Predicament of Mankind.* New York: Universe Books.

Meadows, D.H., J. Randers, and D.L. Meadows. 2004. *Limits to Growth: The 30-Year Update.* White River Junction, VT: Chelsea Green Publishing Company.

Mildner, S-A, G. Lauster, and L. Boeckelmann. 2013. "Scarce Metals and Minerals as Factors of Risk: How to Handle Criticality." Lecture notes in logistics. *Supply Chain Safety Management,* 51–58.

Minerals Education Coalition. 2016. *Estimated Minerals Consumption Over the Lifetime of Every American Born Today.* Society for Mining, Metallurgy, and Exploration.

National Commission on Materials Policy (NCOMP). 1973. *Material Needs and the Environment Today and Tomorrow.* Washington, D.C., June.

National Mining Association. 2012. "Minerals: America's Strength," Accessed May 16, 2015.

Organization for Economic Cooperation and Development (OECD). 2015. *Material Resources, Productivity, and the Environment.*

Parthemore, C. 2011. *Critical Minerals: Growing Demands, Rising Tensions.* Center for a New American Security.

Perry, M. 2012. "Dangerous Dependence: US Increasingly Beholden to Imported Raw Material." *Detroit News,* April 5.

President's Materials Policy Commission. 1952. *Resources for Freedom, Volume I-Foundations for Growth and Security.* U.S. Government Printing Office, Washington, DC.

PricewaterhouseCoopers Accountants N.V. 2011. "Minerals and Metals Scarcity in Manufacturing: The Ticking Time Bomb." *Sustainable Materials Management.*

Rio Tinto. 2010. *Fixed Income Investors Update.* March 31.

Ripley, E., R. Redmann, and A. Crowder. 1996. *Environmental Effects of Mining.* Delray Beach, Florida: St. Lucie Press.

RISI. 2015. *Lumber Dashboard.* January.

Rogers, E., I. Stovall, L. Jones, R. Chabay, E. Kean, and S. Smith. 2000. *Fundamentals of Chemistry.* Falcon Software, Inc.

Steel Recycling Institute. 2014. *2013 Steel Recycling Rates.*

Tilton, J. 2003. *On Borrowed Time? Assessing the Threat of Mineral Depletion,* 101–103. Washington D. C.: Resources for the Future.

Tilton, J. and Guzmán, J. 2016. *Mineral Economics and Policy.* Washington D.C., Resources for the Future.

United Nations University. 2013. *Critical Minerals for the EU Economy: Foresight to 2030.* Institute on Comparative Regional Integration Studies.

U.S. Department of Energy. 2013. *Rare Earth Metals.* Ames Laboratory.

U.S. Geological Survey. 2016. *Mineral Commodity Summaries 2016:* U.S. Geological Survey.

Wellington, T., and T. Mason. 2014. *Effects of Production and Consumption on Worldwide Mineral Supply.* University of Tennessee/Oak Ridge National Laboratory.

World Economic Forum. 2014a. *Global Risks 2014.*

World Economic Forum. 2014b. *The Future Availability of Natural Resources: A New Paradigm for Global Resource Availability.* November.

Yellishetty, M., G. Mudd, P. Ranjith, and A.Tharumarajah. 2011. "Environmental Life-Cycle Comparisons of Steel Production and Recycling: Sustainability Issues, Problems, and Prospects." *Environmental Science and Policy* 14(2011): 650–663.

CHAPTER 8

Bio-Resources and Products

*For the first time in 60 years, the carbohydrate economy
is back on the public policy agenda, changing the very
material foundation of industrial economies.*

—DAVID MORRIS

Bio-products are derived from biological resources—principally plants. They come from agricultural crops and crop residues, natural forests, tree plantations, algae, byproducts of food processing, and animal wastes. Biological resources are renewable as long as they are managed sustainably, a reality that has led to heightened interest in such resources in recent decades. Tremendous potential exists for greater reliance on renewables as sources of heat, power, and liquid fuels, as construction materials, and as chemical building blocks for a wide range of consumer products. The terms "bio-economy" and "carbohydrate economy" are often employed in referring to the economic implications of a shift toward expanded use of biological raw materials.

Wood is by far the most used non-food bio-resource. Among the economically less developed countries, the primary use of wood is as a fuel. Forty percent of the world's population is reliant on wood, charcoal, and other forms of biomass as their primary source of non-transportation energy. In fact, so much wood is used for heating and cooking that fuelwood ranks as the number one use of wood worldwide. Other major uses include paper, furniture, and building materials. Globally, at least 1.3 billion people, or 18 percent of the world's population, live in wood houses.

In some countries wood has become a sought-after source of energy for commercial electricity generation, and production of heat and steam for use in industrial plants and commercial enterprises. Wood has long been used in northern Europe to fuel district heating systems. Relatively recently, research has been mounted to develop technologies for converting wood to industrial chemicals, plastics, liquid fuels, and other products now derived from petroleum.

Trees, although they are by far the greatest contributor to bio-products, are not the only bio-resource. Agriculture has for many decades been a source of products other than food. Products include fiber used in making textiles, limited quantities of paper, pharmaceuticals, and specialty chemicals. More recently, agricultural products have also included ag-fiber based composite panels, renewable plastics and other biopolymers, liquid fuels such as ethanol and biodiesel, and other energy products.

To some, renewal of interest in bio-resources signals a sea change in social thinking. Visionary David Morris briefly outlined the history and future of reliance on biomaterials in the fascinating article "The Once and Future Carbohydrate Economy." In it he says in part:

> Less than 200 years ago, industrializing societies were carbo-hydrate economies. In 1820, Americans used two tons of vege-tables for every one ton of minerals. Plants were the primary raw material in the production of dyes, chemicals, paints, inks, solvents, construction materials, even energy. . . . Then we shifted to a fossil-fuel based economy based on low crude oil prices. By 1920 the nation had reversed the vegetable-mineral ratio, using two tons of minerals for every one ton of vegetables.

Morris described how interest in biomaterials again surfaced in the course of World War II, pointing out that:

> In 1941, when Japan cut off access to Asia's rubber plantations, the United States launched a crash synthetic rubber program. Washington drafted into service both the nation's oil refineries and breweries. In 1943, most of America's synthetic rubber was made from ethanol. By 1945, the United States produced over

600 million gallons of ethanol, a level not again attained until the mid-1980s. A small amount of ethanol was made from wood.

. . . Meanwhile, the carbohydrate economy was featured in the popular press and newsreels, reporting on such sensational developments as Henry Ford's biological car. The body of the 1941 demonstration vehicle consisted of a variety of plant fibers, including hemp. The dashboard, wheel, and seat covers were made from soy protein. The tires were made from goldenrod, bred by Thomas Edison on his urban farm in Fort Myers, Florida. The tank was filled with corn-derived ethanol.

. . . Now, for the first time in 60 years, the carbohydrate economy is back on the public policy agenda, changing the very material foundation of industrial economies. Whether and how we affect that change can profoundly affect the future of our natural environment, our rural economies, agriculture, and world trade. It is an exciting historical opportunity, but one we should approach with deliberation and foresight.

Biomaterials and Petroleum

For now, the vast majority of industrial chemicals used to make a wide range of products are derived from petroleum. While a barrel of oil is typically thought of in terms of gasoline, diesel fuel and heating oil, crude oil is also the source of a range of other common products including asphalt, motor oil, lubricants, most plastics, synthetic fibers, refrigerants, propellants, and wax. It is also used also in making other products ranging from pharmaceuticals, food products, and cosmetics to DVDs, tennis rackets, and contact lenses.

The resurgence in interest in biomaterials as sources of energy and materials is based on two major factors:

1. Realization that petroleum reserves are finite.
2. Concern about carbon emissions associated with extraction, processing, and combustion of fossil fuels.

At the point that petroleum becomes less available and increasingly expensive, bio-resources can be used as an alternative source of not only energy, but also industrial chemicals, solvents, pharmaceuticals, and plastics that can supplement or take the place of similar products now obtained from petroleum. Research is also underway to develop economical pathways for conversion of wood to liquid fuels, including ethanol and jet fuel. Some technologies are already in use, while others are in various stages of development.

Products Obtained from Refining a Barrel of Petroleum (U.S. Average Yields, 2014)

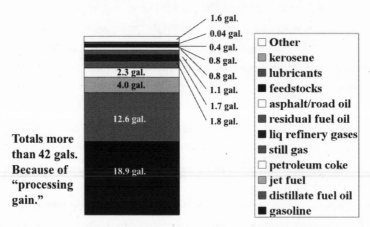

Source: U.S. Energy Information Administration (2015).

The Promise of Agriculturally Derived Materials

The potential contribution of biomass to primary energy and industrial chemicals production within the U.S. has been extensively studied over the past several decades. The potential is seen as far greater than the current level. For instance, it was determined in the mid-1990s that there were about 392 million acres (192 million ha) in the continental U.S. that were not being used for food production and that had the capacity of producing significant quantities of biomass without the need for irrigation. Of these, some 55 million acres (22 million ha) were identified as available and having high potential for production of energy crops such as switchgrass, reed canary grass, poplar, eucalyptus, and other species.

In 2011 the U.S. Department of Energy estimated that 910 million dry tons of biomass crops and agricultural residues could be produced annually in the United States. Agricultural residue estimates include only those residues produced in excess of that needed for conservation till-age which could be removed annually from land used for agricultural production. Another 100 million dry tons of woody biomass could be sustainably removed annually from the nation's forest lands and gleaned from current waste streams; a part of the woody biomass would come from non-commercial forest thinnings conducted for the purpose of re-ducing wildfire danger. Estimated biomass availability does not include resources that are currently being used, such as corn grain and forest products industry mill residues.

Biomass is not the only potential source of bioenergy. Options in-clude such things as methane gas from old landfill sites or from manure collected at feedlot or dairy operations, organic materials recovered from wastewater, and municipal solid waste. The volume of manure alone, con-sidering only the volume in excess of that which can be applied for on-farm soil improvement, is estimated at 12–20 million dry tons annually.

Overall, potential contributions of biomass to U.S. energy and indus-trial chemical supplies are estimated to be substantial. Estimates indicate that by 2030 biomass could supply five percent of the nation's electrical power, 20 percent of its transportation fuels, and 25 percent of its indus-trial chemicals and feedstocks, equivalent to 30 percent of 2010 petro-leum consumption.

Potential use of agricultural lands for increased production of energy and chemicals is controversial. A major concern is displacement of food crops at a time when feeding the growing global population is increasingly problematic. The current practice of producing ethanol from corn starch and biodiesel from soy and canola oil exemplify concerns. However, de-velopment of second generation biofuels that will allow production of fuels from plant stalks rather than vegetable content will alleviate some of that concern. In addition, as noted in the following chapter, yields of what are known as second generation biofuels will be far greater per ton of biomass input than are currently being realized.

Studies in the European Union have shown potential for providing 11 percent of energy supplies and 5–6 percent of transportation fuels from biomass. Just considering the transportation sector, estimates indicate that over 13 billion gallons of fossil-based transportation fuels could be replaced by biomass-derived fuels. Similar evaluations are being done all over the world. Although studies of potential to this point have not been as extensive as those in the U.S. and EU, virtually all indicate the possibility of increasing production of energy and essential chemicals from biomass resources.

Wood—A Premiere Bio-Material

In a number of economically advanced countries, such as the U. S., Canada, Japan, New Zealand, Australia, Sweden, and Finland, the vast majority of wood consumed is used to produce tangible products. The primary use is as a building material, with wood accounting for a large portion of residential construction and a growing segment of non-residential building projects. In the United States and Canada more than 85 percent of residential homes are wood framed. In Europe, wood is typically used in framing roofs and for interior applications, although structural use of wood is increasing across the continent. Most paper is made of wood, something that is true for the world as a whole. In fact, wood accounts for over 95 percent of global pulp production.

Most of the annual wood harvest in the United States goes into paper, solid wood products manufacturing, and energy products. Paper is made from both virgin and recovered fiber. Despite increased recycling of waste paper, two-thirds of paper production still comes from trees harvested as pulpwood, wood chips, and other residues obtained from sawmill trimmings.

Solid wood products include a wide range of wood construction materials that are beginning to find new applications in structures of all kinds. A casual observer of change in building materials technology and construction practices in the United States could easily have missed the dramatic shift toward engineered wood products over the past several decades. Throughout the 20th century, steel and concrete dominated the tall buildings scene while wood framing dominated single-family and low-rise residential construction. During this same period, large wooden columns and beams, made by gluing smaller pieces of wood together with the grain of all pieces parallel to one another—today commonly known as "glulam beams"—became increasingly common in construction of church buildings and other types of structures. However, development of new types of engineered wood products in North America in the 1980s and 1990s significantly increased design options for wood structures. These developments, coupled with interest in more economical construction, led to an increase in the number of four- and five-story wood-framed buildings around the beginning of the 21st century, a trend that continues today.

Acceptance of taller wood-framed buildings is now overshadowed by a new development—one that has its roots in Europe. Almost overnight, the height potential for modern wood buildings has increased dramatically. The newest development is introduction of cross-laminated timber (CLT) which is made of a number of layers of lumber, glued together with the grain of alternate layers laid at right angles to one another, much like the veneers of plywood. CLT panels can be as large as 20 inches thick x 10 feet x 60 feet (0.5 x 6 x 18 meters). CLT combines the advantages of large size load-bearing panels, stability, fire resistance, long-term carbon storage, and renewability. Engineers and architects soon discovered that through the use of large CLT panels, wood buildings could be constructed to previously unimagined heights. Within a period of only six years a number of number of wood buildings were constructed around the world ranging in height from six to 14 stories, and it has been demonstrated that wood structures as tall as 40 stories are technically feasible. Expansion of this concept is likely.

A small percentage of the annual harvest in developed countries is used for firewood, and an increasing portion goes into the production of fuel pellets for domestic use and export, to be used in commercial power generation and institutional and home heating. Use of wood for energy production, and especially of fuel pellets for export, has triggered concerns about impacts on forests, although studies suggest that pellet markets are actually likely to encourage retention or expansion of forested areas in the U.S.

Forests and Wood Resources

About 30 percent of the world's land area is covered by forests, amounting to around 10 billion acres (4 billion hectares). The near and long-term availability of wood resources is obviously dependent upon forest retention and sustainable management of forest resources. In this regard, significant differences exist between various regions. While forest area is declining in some parts of the world, it is stable or increasing in others. Specifically, forest area is declining in South and Central America, Southeast Asia, and Africa, and is stable or increasing in North America, Europe, China, India, Siberia, and the Caribbean.

Changes in Forest Area by World Region, 1990–2015 (Percent gain or loss)

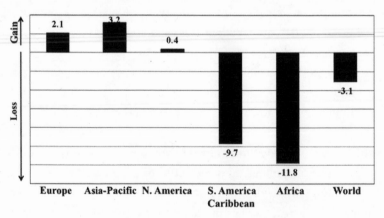

Source: Keenan et al. (2015).

Forest retention is currently most problematic in the world's tropical regions. Significant deforestation is occurring in those areas, driven by population growth and increasing consumption both locally and globally. Losses are largely attributable to shifting agriculture and conversion to large and small scale permanent agriculture, including conversion of natural forests to natural rubber and palm oil plantations. The term shifting agriculture is used to describe the practice of clearing an area of forest, followed by cultivation for a few years, and then abandonment and repeat of the process in a new area as soil fertility wanes. Illegal and unsustainable logging and collection of firewood are major factors in forest degradation, often the first step toward deforestation.

Main Causes of Tropical Deforestation by World Region, 2000–2010

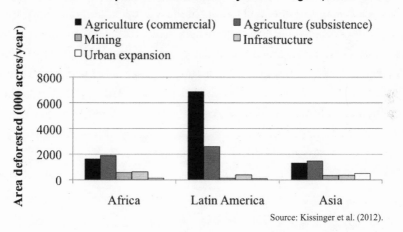

Source: Kissinger et al. (2012).

Main Causes of Tropical Forest Degradation by World Region, 2000–2010.

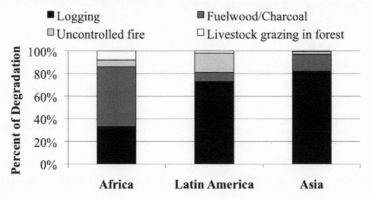

Source: Kissinger et al. (2012).

In Africa, where forest loss as a percent of forest reserves is currently greatest, over 90 percent of annual wood removals are for fuelwood. In Asia and South America fuelwood accounts for 75 and 50 percent of removals, respectively. In Oceania, North America, and Europe it is a much different story, with the portion of annual harvest used for wood fuel at 0, 10, and 25 percent, respectively.

The forest situation in North America is far different from that in the tropics, a reality often obscured by misconceptions. For instance, surveys of thousands of college students and others over the past 20+ years have shown a pervasive belief that forests of the United States are disappearing. Fortunately, reality does not mesh with that perception. In fact, the forest land area has remained remarkably constant over the past 100 years. Timberland (forests that are available for periodic timber harvest) has similarly remained largely unchanged over an extended period. Timberland makes up about two-thirds of the forest area of the United States.

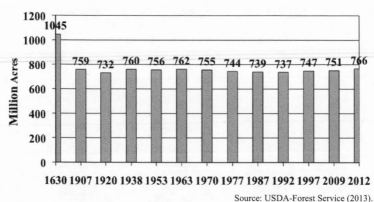

Trends in U.S. Forestland Area, 1630–2012

Source: USDA-Forest Service (2013).

The extent of forests in the United States in 2016 remained about the same as in 2012 and in 1900, and about two-thirds of what existed at the time of European settlement. The bad news in this story is that the population of the U.S. quadrupled in the 116 year period 1900–2016, from 76 million to about 323 million, with the result that the forest area *per capita* declined accordingly—from about 10 acres (4 ha) in 1900 to one-fourth of that (2.4 acres, or 1 ha) in 2016 (see page 75). This trend is likely to continue

into the foreseeable future with continued population growth, urbanization, and infrastructure expansion, ensuring that conflicts over forest use
and management will persist and intensify in the decades ahead.

An unchanging forest area does not mean that forests cover exactly
the same geographic area as in decades past. Nor does it mean that the
structure of forests is unchanged; today there are fewer larger, old-growth
trees, and greater numbers of smaller, younger trees than in times past.
Forest cover in the U.S. and worldwide is dynamic, changing in response
to a number of events and trends. Maintaining the current extent of forest in the United States throughout the current century will likely be a
significant challenge given that the population is expected to increase
substantially, with growth concentrated in several heavily forested states.
Urbanization is likely to increasingly displace forests.

Another widely-held misconception is that more timber is harvested
each year than is grown. Despite rising population, wood consumption, and
harvest levels, net forest growth in the U.S. has exceeded removals for at
least seven consecutive decades, with an increasing margin between growth
and removals. In 2015 the rate of tree growth and wood formation was more
than double the rate of removals. Because the volume of wood added annually through new growth exceeds the volume of wood removed, the volume
of trees (and of wood) in U.S. forests is rising. Trees within the one-third of
U.S. forests that are off-limits to periodic harvesting, such as forests within
wilderness areas, are not counted when determining timber volume.

Net Annual Growth vs. Annual Removals in U.S. Forests (2012)

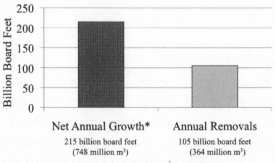

Net Annual Growth*	Annual Removals
215 billion board feet	105 billion board feet
(748 million m³)	(364 million m³)

* Net growth is equal to the volume of all new wood formation minus losses
due to natural mortality, fire, insects, disease, and all other factors.

Source: Oswalt et al. (2014)

An issue that has received increasing attention in the U.S. in recent decades is more frequent occurrence of catastrophic forest fires in the western
states. Due, in part, to a warming climate and drought, the situation is accentuated by overly-dense forest stands, especially within federally-owned
National Forests. A long history of fire suppression combined with sharp
declines in harvest activity within these forests over the past several decades
has created vast areas of high biomass concentration that increase chances
of disease and insect infestation and, if ignited, burning with great intensity.
Without a change in policy, a continuation of recent trends is likely. Neglect
of the federal forests has led to a situation in which annual tree mortality in
recent years has approached 90 percent of net annual growth, while only
slightly over 11 percent of net growth has been harvested. That translates to
massive accumulation of combustible biomass. Comparing these statistics
to those prior to the recession years of 2007–2010 shows greater harvest
levels prior to the economic downturn, but with removals still only 25 percent of net growth. A policy of removing more timber from the forest each
year than is created through annual growth is clearly not sustainable. It is
becoming increasingly clear that it is also not sustainable for annual growth
to far exceed rates of annual removals.

Canada has 91 percent of the forest cover that existed prior to European
settlement. As in the U.S., there is a long history of greater net growth
than removals nationwide. However, a devastating mountain pine beetle
outbreak in the western provinces that began in the early 1990s, and
which continued for several decades, greatly reduced the net increment.
Lower harvest levels over an extended period will be needed to return to
a pattern of positive net growth.

Renewability of Forests and Wood

Fortunately, forests are renewable. Unfortunately, that fact is often not
fully appreciated. An example of forest renewability is provided by the
60-year period ending in 2010, during which over 847 *billion* cubic feet
of timber were harvested from U.S. forests, a volume approximately
equivalent to a bulk pile of wood (no spaces) measuring 1 mile x 1 mile x

5.75 miles in height! This wood was used in building some 90 million homes and in producing countless other products. And what was the impact on domestic forests? Because forests are renewable, the volume of wood within domestic forests *increased* by more than 50 percent during those 60 years. Accompanying the massive use of wood was an increase in the volume of carbon stored within U.S. forests, long-term storage of billions of tons of carbon within residential structures and other buildings, and avoidance of even greater quantities of carbon through use of wood rather than other more fossil-energy intensive products.

The differences between renewable and non-renewable materials are fundamental and dramatic. These differences are sometimes overlooked or discounted in discussions of environmental policy. Maintaining access to a versatile raw material that, if properly managed, will not only maintain but actually increase its volume is a gift that should not be taken lightly.

Wood—Could More Be Harvested?

In some parts of the world, where forest cover is stable and growth rates are well above harvest levels, forest harvesting could be increased without depleting the resource. In fact, there is considerable evidence that the number of trees and volume of wood stored in U.S. forests are far above historic norms. U.S. forests may, therefore, actually be suffering from a level of harvest that is too low. Increasing harvest levels could reduce the need for imports and decrease balance of trade deficits, while at the same time improving forest conditions over large geographic areas. In addition to the potential for greater harvest in the U.S., the vast (mostly hardwood) forests of Brazil and other parts of South America could conceivably support a larger sustainable harvest.

There is, however, significant public opposition to increasing harvest levels worldwide, and especially in the most economically developed countries. Consequently, the near-term potential for increased harvesting in natural forests is effectively limited to relatively few areas of the world. These include Siberia, the far eastern region of Russia, and several eastern European nations. The extent to which global and regional economic constraints, and sometimes ill-advised environmental concerns, limit future increases in harvest remains to be seen. In the meantime, establishment of fast-growing industrial wood plantations is occurring all over the world.

The Role of Forest Plantations

Forest plantations generally produce much more wood per geographic area than natural forests because they are usually established on highly productive sites, intensive silviculture (sometimes including fertilization) is practiced, and genetically-selected growing stock is used. Wood production in these plantations ranges from two to five times that of an equal area of natural forest, and in some cases can be as much as 10 times greater than surrounding forests. Plantations will clearly play a significant role, and perhaps even a dominant role, in providing future wood supplies.

The area of plantations worldwide was estimated at 687 million acres (278 million ha) in 2015, comprising about seven percent of the total forest area worldwide. Over the period 2010–2015, plantations increased at a rate of about 54 million acres (22 million ha) annually. Almost half of the world's forest plantations are currently located in Asia, primarily East Asia including China, where plantations account for over 35 percent of its forests. Globally, about three-fourths of forest plantations are located in Asia, Europe, and North America.

Plantation forests have been established around the world to provide industrial wood (industrial plantations), fuelwood, and natural rubber. Planted forests are sometimes also established to provide soil and water conservation and protection from wind erosion.

Fast-growing industrial wood plantations cover 134 million acres, accounting for a little over 20 percent of the total plantation area and just 1.3 percent of total forest coverage worldwide. Yet, these plantation forests satisfy about one-third of global consumption of industrial wood, a number that is expected to increase over time.

In addition, there are approximately 20 million acres of fuelwood plantations in the world, with most of these in Asia. The gathering of wood for use as fuel is a cause of increasing concern. In fact, this use is growing faster than world population. Specifically, consumption of fuel-wood increased 250 percent between 1960 and 2012, a period in which world population increased by 90 percent. Plantations of fast-growing tree species provide a way to reduce pressures on natural forests while providing a source of sustainable fuelwood supply.

Despite the great promise of intensively-managed plantations for pro-viding needed volumes of wood, considerable controversy exists. One of the controversial aspects of plantations is the clearing of natural forests to make way for plantations. Controversy is also rooted in perceptions of low biodiversity of plant and animal species in planted forests compared to natural forests, common use of non-native tree species in plantation es-tablishment, long-term forest health risks associated with limited genetic variation in planting stock, and land ownership issues.

Extending Wood Resources through Recycling

There remains significant opportunity for greater recovery and recycling of wood resources globally. There is potential for greater recovery and recycling of paper, and particularly large possibility for greater recovery, reuse, and recycling of solid wood materials.

More than 400 million metric tons of paper and cardboard are pro-duced worldwide every year, with more than half coming from recovered sources. At current levels of technology, paper fibers can be recycled 4–9 times before they are sufficiently degraded to be lost to reuse. This means that an upper limit to reuse is about 80 percent, as at least 11 to 25 percent of virgin fiber is needed in the fiber cycle.

In the U.S. about two-thirds of waste paper generated is recovered for recycling. Virtually all of this is used in paper and paperboard manufac-turing, although a considerable portion goes to production mills outside the United States. The recycled content of domestically manufactured paper was 37 percent in 2013.

Global statistics regarding recycling of non-paper wood products are not available. Current data for the U.S. indicate that the overall recycled content of domestically produced wood building products, such as lumber and other structural materials, is in the 10–11 percent range, a figure that shows considerable room for improvement. A 2014 estimate indicated that the U.S. generates about 70 million tons of wood debris annually. This is made up of municipal solid waste and construction and demolition discards. Of that, only about 28 million tons is available for recovery. Municipal solid waste includes yard waste, while construction and demolition waste is generally limited to various forms and pieces of building products. Various efforts are underway to recover a greater portion of discards, both through increasing the frequency of deconstruction, rather than demolition, and reuse of building components, and increasing recovery of discards for either reuse or energy capture. Increased fabrication of building components in factory settings is also being eyed as a way to reduce job-site waste and increase materials use efficiency.

Import Dependence

During the period 1950–2012, the volume of timber in forests of the U. S. increased by over 57 percent at the same time that the population grew by 96 percent. One reason that the timber volume has increased despite expanding wood consumption is that the U.S. has increasingly become dependent on imported wood and softwood in particular. For the past several decades, net imports of softwood lumber and products (i.e., those products used primarily in construction) have ranged from about 23–35 percent, with Canada supplying the vast majority of imported wood.

Because Canada manages its forest resources responsibly, the primary environmental impacts of U.S. import reliance on Canada occurs in other countries. Canada, if not supplying timber to the U.S., could be exporting to a greater extent than it does to China, Japan, Korea, Taiwan and other countries. These countries go elsewhere to obtain wood supplies, often from regions such as the Russian Far East where logging is still exploitive.

Biomaterials and the Environment

Wood is strong, lightweight, easily cut-shaped-and fastened, durable, naturally beautiful, and recyclable. It is formed within growing trees using solar energy, and can be transformed into useful products with the addition of relatively little additional energy. Moreover, because wood in its various forms has many and varied uses—from simple or highly engineered structural materials or a source of energy, to papermaking fiber and feedstock for industrial chemicals production—there is essentially zero waste in wood products manufacture. And, wood is both abundant and renewable as long as it is harvested from responsibly managed forests.

Not surprisingly, environmental impacts from production of wood products are almost always lower than those associated with similar products made of substitute materials. Differences are often substantial, and are readily seen in comparisons of buildings. When otherwise identical buildings made of wood, steel, and concrete are compared using life cycle assessment, wood structures invariably exhibit strikingly lower impacts across a wide range of impact measures. For example, life cycle assessments consistently show construction of wood buildings to require 20 to 40 percent less energy, and 25–50 percent less fossil energy than the same buildings constructed primarily of steel or reinforced concrete. Emissions linked to energy use are likewise lower. In addition, since wood captures carbon dioxide as it grows, incorporating it into wood, that carbon is retained within the wood structure for as long as the wood exists, providing long term storage of carbon in wood products.

Public opinion surveys have shown that many people view wood as having lower durability than steel or concrete. Yet, an extensive study of the performance of actual buildings has shown no difference in long-term durability of structures made of various structural materials. And, while wood of small cross-section will readily burn, wood in large sizes is far more resistant to fire and more resistant to catastrophic failure than steel. Realistically, buildings are seldom constructed of a single material. A commercial or multifamily residential building may, for instance, have a concrete foundation and ground floor, five stories of wood above that, and metal fasteners or reinforcing rods throughout. Yet, careful analysis of various components consistently shows lower impacts for components made of wood than of functionally equivalent components made of other materials. In view of this, and the fact that wood is renewable, an obvious conclusion is that wood should preferentially be used in construction or other applications wherever it makes sense to do so.

While there are obviously limits to wood availability, there is today (2017) tremendous potential for shifting to use of wood as a replacement for many of the non-renewable materials used today. In the near to mid-term, wood will play an increasingly important role in construction of commercial buildings (including tall buildings), production of energy and industrial chemicals, and even for such applications as electronics components, batteries, and high impact glass.

References

Barua, S., and P. Lehtonen. 2014. "The Great Plantation Expansion." *ITTO Tropical Forest Update* 22/3.

Bentsen, N., and C. Felby. 2012. "Biomass for Energy in the European Union—A Review of Bioenergy Resource Assessments." *Biotechnology for Biofuels* 2012, 5:25.

Birdsey, R., K. Pregitzer, and A. Lucier. 2006. "Forest Carbon Management in the United States: 1600–2100." *Journal of Environmental Quality* 35:1461–1469.

Bowyer, J. 1995. "U.S. Forests and Forest Products: Fact vs. Perception." *Forest Products Journal* 45(11/12): 17–24.

Bureau of International Recycling. 2015. "Paper," Accessed December 10, 2015.

Butler, R. 2012. "Fuelwood Harvesting as a Threat to Rainforests." Last updated July 31, 2012.

Dawkins, H. 1958. *Silvicultural Research Plan, First Revision, Period 1959 to 1963 Inclusive.* Forest Department, Entebbe, Uganda.

Food and Agricultural Organization of the United Nations (FAO). 2010. *Global Forest Resources Assessment 2010.* FAO Forestry Paper No. 163.

Food and Agricultural Organization of the United Nations. 2014. "Putting People at the Centre of Forest Policies," FAO News Release, June 23.

Food and Agricultural Organization of the United Nations (FAO). 2015. *Global Forest Resources Assessment 2015: How have the world's forests changed?* FAO Forestry Paper.

Frederick, K., and R. Sedjo, eds. 1991. *America's Renewable Resources: Historical Trends and Current Challenges.* Washington, DC: Resources for the Future.

Green, M. and Taggert, J. 2017. Tall Wood Buildings: Design, Construction and Performance. Birkhauser Verlag, GmbH.

Ince, P. 2010. "Global Sustainable Timber Supply and Demand." In *Sustainable Development in the Forest Products Industry,* edited by R. Rowell, F. Caldeira, and J. Rowell, 29–41. Porta, Portugal: Universidade Fernando Pessoa.

Keenan, R., Reams, G., Achard, F., deFreitas, J., Grainger, A. and Lindquist, E. 2015. Dynamics of Global Forest Area: Results from the FAO Global Forest Resources Assessment 2015. Forest Ecology and Management 352 (2015): 9–20.

Kissinger, G., M. Herold, and V. De Sy. 2012. *Drivers of Deforestation and Forest Degradation: A Synthesis Report for REDD+ Policymakers.* Lexeme Consulting, Vancouver Canada, August.

Lovejoy, T. 1990. *Consensus Statement on Commercial Forestry Sustained Yield Management and Tropical Forests.* Smithsonian Institution.

MacCleery, D. 2011. *American Forests: A History of Resiliency and Recovery.* Durham, NC: Forest History Society.

Morris, D. 2006. "The Once and Future Carbohydrate Economy." *The American Prospect,* March 20.

O'Connor, J. 2004. Survey of Actual Service Lives for North American Buildings. Woodframe Housing Durability and Disaster Issues Conference. Las Vegas, October.

Oswalt, S., W.B. Smith, P. Miles, and S. Pugh. 2014. *Forest Resources of the United States, 2012.* Gen. Tech. Rep. WO-91. Washington, DC: U.S. Department of Agriculture, Forest Service, Washington Office.

Smith, W., P. Miles, C. Perry, and S. Pugh. 2008. *Forest Resources of the United States, 2007.* USDA-Forest Service. General Technical Report WO-78.

Tuskan, G., M. Downing, and L. Wright. 1994. "Current Status and Future Directions for the U.S. Department of Energy's Short-Rotation Woody Crop Research." In *Proceedings: Mechanization in Short-Rotation, Intensive Culture (SRIC) Forestry,* edited by B. Stokes, and T. McDonald. Mobile, Alabama, March 1–3.

USDA-Forest Service. 2001. *U.S. Forest Facts and Historical Trends.* FS-696-M.

U.S. Department of Energy. 2011. *U.S. Billion-Ton Update: Biomass Supply for a Bioenergy and Bioproducts Industry.* R.D. Perlack and B.J. Stokes (Lead authors), ORNL/TM-2011/224. Oak Ridge National Laboratory, Oak Ridge, TN.

Whitmore, L. 1999. "The Social and Environmental Importance of Forest Plantations with Emphasis on Latin America." *J. Tropical Forest Science* 11(1): 255–269.

World Bank. 2008. *Introduction: Opportunities and Challenges in the Forest Sector,* 2.

CHAPTER 9

Energy

The fact remains that, if the supply of energy failed,
modern civilization would come to an end as abruptly
as does the music of an organ deprived of wind.

—FREDERICK SODDY

Energy, and lots of it, is essential to maintenance of life styles in the most economically developed nations. Profligate energy consumption is also a hallmark of emerging middle classes in economies around the world. At the same time, access to energy is a goal of millions more. In early 2015 some 1.3 billion people around the globe, equivalent to more than four times the population of the United States, did not yet have access to electricity.

Most of the world's largest energy consumers are also net importers of energy, and often on a massive scale. Sourcing energy has long been a cause of unease among world leaders, and a key consideration behind Japan's entry into World War II. Concerns about adequacy of energy supplies have increased over the past quarter century as economic gains in Asia, and particularly China, have greatly accelerated global energy consumption. Before that time, just as with non-fuel minerals, the most economically developed nations faced little competition for the world's energy resources. That this is no longer the case is exemplified by China which was a net importer of fewer than 10,000 barrels of oil daily in 1970. Net imports were almost 700 times that in 2014.

Globally, per capita consumption of primary energy increased by a factor of nine times between 1850 and 2010. This, combined with a 5.5-fold increase in population over that time period, translates to a 49-fold increase in annual energy consumption. Just shy of 90 percent of this increase in consumption was fossil fuels, for which the rise in use was nothing short of stunning. As of 2010, about 50 percent of all fossil fuels consumed in all of history had been consumed in just the previous 25-year period.

Fossil fuels—petroleum, natural gas, and coal—are overwhelmingly the dominant sources of energy in the early 21st century. In response to rising energy consumption and competition for energy supplies, energy producers have introduced new technologies to increase production, creating excitement about potential for substantial increases in supplies of petroleum and natural gas. At the same time, new extraction techniques have raised concerns about impacts on freshwater supplies. Meanwhile, the world's principal sources of energy have become recognized as the primary contributor to rising levels of atmospheric carbon dioxide and changes in the earth's climate.

Energy, and its future supply, presents a complex puzzle for humankind. Timely solutions, or a lack of thereof, will likely have an enormous impact on civilization for many generations to come.

Sources of Energy

Petroleum (oil) is the most important energy source. Coal, natural gas, hydropower, nuclear, and renewables are all significant sources of energy. Renewable energy worldwide is dominantly produced from biomass (primarily wood). Household use of biomass for home heating and cooking accounts for over one-half of the wood harvested globally each year, with virtually all such use occurring within the developing countries. Wood and other biomass is also increasingly used in developed and developing countries alike for production of electricity and liquid fuels such as ethanol and biodiesel, and as a source of direct or steam heat for factories and commercial buildings. Overall, wood and biomass accounted for 10.3 percent of the total energy supply in 2014. Other sources of renewable energy (and the percent of the total global energy supply produced using them) include nuclear (4.8 percent of energy overall and 12.1 percent of electricity), hydroelectric (2.4 percent), and all other renewables, including geothermal, solar, and wind energy (1.4 percent). Nuclear, hydro, and wind energy together provided almost one-fifth of electrical energy globally in 2014.

**Fuel Shares of World Energy Supply, 1973 and 2014
(Million tons of oil equivalent – MTOE)**

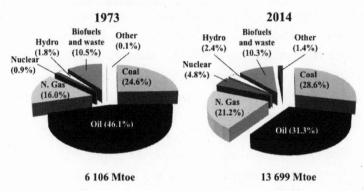

Source: International Energy Agency (2016).

**Primary Energy Use in the United States by Source, 2015
(Quadrillion Btu and percent of total).**

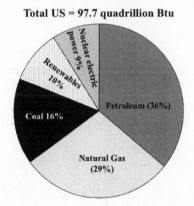

Source: U.S. Department of Energy, EIA (2016).

In the United States and other developed countries, the energy picture is similar to that of the world at large, except that little wood is used for heating and cooking. Using the U.S. as an example, fossil fuels are again the dominant energy sources, with petroleum the largest single source. As of 2015, renewables accounted for 10 percent of primary energy consumption, with biomass-derived energy (4.9 percent), hydroelectric (2.5 percent), and wind (1.9 percent) the leading sources.

Energy from biomass includes liquid biofuels, such as ethanol (from corn) and biodiesel (from soybeans), and electricity generated primarily from wood. Significant volumes of wood are also used by the domestic forest products industry to produce process heat. As is the case globally, other renewables, including geothermal and solar make up a small percentage of energy consumption, together totaling a little more than one-half of one percent. Nuclear energy provides 9 percent of total primary energy and 20 percent of electricity.

The makeup of energy supplies in other developed nations is similar to that of the U.S. For example, petroleum is the dominant energy source in the European Union, Canada, Australia, and Japan, with coal and natural gas either second or third leading sources of energy in all of these countries. Nuclear energy accounts for a greater portion of supply in the EU (14.4 percent), and renewables contribute more prominently to energy supplies in the EU (11.6 percent) and in Canada (18.3 percent), than in the U.S.

Abundance and Availability of Fossil Fuels

Hydraulic fracturing (also known as fracking), the ability to extract petroleum and natural gas from shale and what are called other "tight formations," has improved the near-term global energy picture. The United States, in particular, has benefitted from this development, with domestic production now increasing after a long period of decline. With abundant reserves of natural gas and coal, the energy security of the U.S. appears assured for many decades into the future. There is one caveat, however. Should society conclude that climate change requires drastic reduction of carbon emissions, fossil fuels, and especially coal and petroleum, could become fuels of last resort.

The global situation relative to fossil fuels looks somewhat less promising. Even though new extraction technologies for oil and natural gas are likely to be implemented globally, the long-term global supply outlook is unlikely to change radically. Rapid consumption growth in Asia, and especially in China and India, is projected to continue. Nonetheless, supplies of oil, natural gas, and coal are expected to be sufficient to meet global consumption needs for the foreseeable future. Reserves life indices (often expressed as the *R/P* or reserves to production ratio) for petroleum, natural gas, and coal were estimated at the end of 2015 at 52.5, 54.16, and 110 years, respectively. As with non-fuel minerals, life index numbers are dynamic, rising as new reserves are discovered or new technologies allow use of previously inaccessible resources, and falling as rates of production and consumption rise.

Petroleum Resources and Long-Term Availability—
A History of Pessimism

For a number of decades society has been worried about the long-term adequacy of fossil fuel reserves, and most specifically petroleum reserves. Concern has centered on "peak oil," a sometimes misunderstood concept. Specifically, peak oil is not the point at which oil is predicted to run out. It is, instead, the point in time when world oil production reaches its highest rate, after which it goes into permanent decline.

The term "peak oil" originated with Dr. M. King Hubbert, a geophysicist who observed that the amount of oil is finite, and that therefore the rate of discovery must eventually reach a maximum and then decline. He noted that this relationship exists for any geographical area and also for the world as a whole. Hubbert warned in a 1949 article in the journal *Science* that the fossil fuel era would be of very short duration. He described the role of fossil energy as "but a pip, rising sharply from zero to a maximum, and almost as sharply declining, and thus representing but a moment in the total of human history." Seven years later, in 1956, he accurately predicted that U.S. oil production would peak in about 1970 and decline thereafter.

Hubbert then turned his attention to global petroleum supplies, predicting in 1969 that world production would peak in the year 2000. His work attracted the attention of a number of others, who over the next 30 years variously predicted the year of peak production within the range 2004 to 2035, with most in the 2010 to 2020 time frame.

The concept of peak oil was derided, discussed, studied, and debated. For the most part, however, peak oil was ignored by world governments. It wasn't until 1998 and another article in *Science* "The Next Oil Crisis Looms Large—and Perhaps Close" that peak oil again appeared on the radar screens of key decision-makers. That article was followed two years later by a comprehensive assessment of global petroleum reserves that concluded that peak oil would occur around 2037. The most optimistic scenario, given no more than a five percent chance of becoming reality, put peak oil at 2047. The most pessimistic put peak production at 2026.

Depiction of Fossil Energy Production and Decline as Described by Hubbert

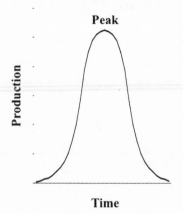

The U.S. government focused on the petroleum issue again in 2005 with the release of another report commissioned by the Department of Energy. What became known as the Hirsch report began with the words "The peaking of oil production presents the U.S. and the world with an unprecedented risk management problem." Among the conclusions were that "prudent risk management requires the planning and implementation of mitigation well before peaking," and that the "the oil peak problem deserves immediate, serious attention."

In early 2007 the National Petroleum Council, an oil and natural gas advisory committee to the Secretary of Energy, released a report focused on petroleum supply. That report spoke about the urgency of today's energy issues; accumulating risks to the supply of reliable, affordable energy; and significant challenges to meeting projected total energy demand. The report also noted that there can be no national energy security without global energy security.

A common thread in all of these reports, and an underlying reason for growing pessimism about long-term petroleum supplies, was sharply rising energy consumption in China and other growing economies, including those of India and other Asian nations. The importance of this phenomenon to world petroleum supplies was emphasized in November 2007, with the release of the International Energy Agency's (IEA) *World Energy Outlook-2007*. An accompanying news release described the consequences of "unfettered growth in global energy demand" for China, India, the OECD and the rest of the world as alarming. According to IEA analyses, net oil imports in China and India combined would jump from 5.4 million barrels per day in 2006 to *19.1 million barrels* per day in 2030—more than the combined imports of the United States and Japan in 2006.

Shortly thereafter, the *Wall Street Journal* carried a front-page news article entitled "Oil Officials See Limit Looming on Production". The article began with the words "A growing number of oil-industry chieftains are endorsing an idea long deemed fringe: The world is approaching a practical limit to the number of barrels of crude oil that can be pumped every day." The article went on to say that evidence is mounting that global production of crude oil may reach a plateau before alternatives are sufficiently developed and "could set the stage of a period marked by energy shortages, high prices and bare-knuckled competition for fuel." A number of executives interviewed in conjunction with the story indicated that they didn't subscribe to the idea that production will be limited by physical supplies of petroleum (the peak oil theory), but rather by a host of other intractable problems.

Now, fast-forward to 2010. In February the U.S. Joint Forces Command published a Joint Operating Environment report which included the sobering observation that even "assuming the most optimistic scenario for improved petroleum production through enhanced recovery means, the development of non-conventional oils (such as oil shale or tar sands) and new discoveries, petroleum production will be hard pressed to meet the expected future [2030] demand of 118 million barrels per day." In addition, the report observed that "By the 2030s, demand [for petroleum] is estimated to be nearly 50 percent greater than today. To meet future global energy demand even assuming more effective conservation measures, the world would need to add roughly the equivalent of Saudi Arabia's current energy production every seven years."

The authors of the Joint Forces report further noted that "the implications for future conflict are ominous, if energy supplies cannot keep up with demand, and should states see the need to militarily secure dwindling energy resources." Like the oil officials cited in the 2007 *Wall Street Journal* report, the Joint Forces Command did not refer to the petroleum problem as one of peak oil—at least over the next decade. They explicitly stated, in fact, that the primary problem for the next ten years is not a lack of petroleum reserves, but rather a shortage of drilling platforms, engineers and refining capacity, a problem that would from their perspective require at least a decade to correct even if corrective action were begun immediately.

The Joint Forces report, while scarcely mentioned in U.S. media, garnered considerable attention overseas. Shortly after its release the German news organization *Spiegel* called attention to a draft report of a German military think tank that had studied the peak oil issue. That report is said to have acknowledged the possibility of peak production around the year 2010 and to have suggested very serious consequences of what was described as a permanent supply crisis, ". . . including a decline in importance of western industrial nations and shifts in the global balance of power."

Also reported were efforts within the government of the United Kingdom to develop a crisis plan to deal with possible shortfalls in energy supply. In the UK, Chatham House-Lloyds had previously released a white paper focused on energy security. The white paper warned of a coming global oil supply crunch, and emphasized the need for businesses to address energy-related risks to supply chains, and particularly enterprises based on just-in-time business models. A key conclusion was that businesses that prepare for and take advantage of the new energy reality would prosper, while failure to do so could be catastrophic.

Release of the 2010 edition of the International Energy Agency's World Energy Outlook caused another stir among energy futures analysts. In a major shift from previous reporting on the matter, the IEA published a graphic of world oil production indicating that a peak in conventional oil production had occurred a year earlier. The IEA report also questioned what were viewed as overly optimistic projections of new crude oil discovery.

World Oil Production by Type in the World Energy Outlook New Policies Scenario, 1990–2035

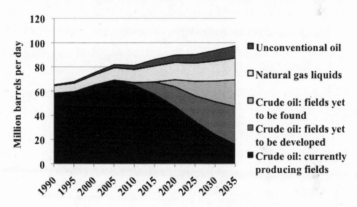

Source: OECD/IIE World Energy Outlook (2010).

Meanwhile, many of the world's energy company executives publicly indicated that peak oil production was at hand. Shell Oil in a 2011 report described the situation this way:

Underlying global demand for energy by 2050 could triple if emerging economies follow historical patterns of development. In broad-brush terms, natural innovation and competition could spur improvements in energy efficiency to moderate underlying

*demand by about 20% over this time. Ordinary rates of supply
growth—taking into account technological, geological, competitive, financial and political realities—could naturally boost
energy production by about 50%. But this still leaves a gap between business-as-usual supply and business-as-usual demand
of around 400 EJ/a—the size of the whole industry in 2000.*

*This gap—this Zone of Uncertainty—will have to be bridged
by some combination of extraordinary demand moderation and
extraordinary production acceleration. So, we must ask: Is this
a Zone of Extraordinary Opportunity or Extraordinary Misery?*

Short-Term Optimism—Long Term Uncertainty

How does the pessimism outlined above square with the new technology
that has unlocked vast oil and gas discoveries in North America and elsewhere? How do such changes affect the future supply outlook?

The most immediate effect is that concerns of a near-term supply
crunch have all but disappeared. Questions remain about the mid- to longer term. As with mineral resources in general, future availability will
depend upon several factors:

- Continued technology development that could make previously inaccessible reserves obtainable.
- Potential limits on application of fracking technology due to concerns
  about impacts on water supplies, seismic activity, and other environmental factors.
- Continued displacement of fossil energy by renewable energy sources.
- Success in limiting increases in consumption through energy
  conservation.
- Potential limitations on the use of coal and other fossil fuels because of
  climate concerns.

The near-term outlook for the U.S. is good, with projections of huge
growth in shale gas production into the foreseeable future and supplies

sufficient to supply all of domestic needs through at least 2040. U.S. shale oil production, on the other hand, is forecast to continue to grow only through 2019, with significant and ongoing decline thereafter. The positive U.S. energy outlook is tempered a bit by anticipated near-term peaking of non-conventional petroleum production. This will have the effect of once again ramping up petroleum imports. Although global petroleum supplies are expected to be adequate for several more decades, adjustments may be necessary to provide for adequate supplies of transportation fuels over the longer term. With petroleum and petroleum derivatives (gasoline, jet fuel, and diesel) the dominant transportation fuels, avoidance of supply problems will likely require greater reliance on biofuels in combination with conversion, where possible, of vehicle energy sources from liquids to natural gas.

Before leaving the topic of U.S. fossil energy potential, it is worth examining well-publicized fracking activity and oil production in the northern plains states. Production is concentrated in what are known as the Bakken and Three Forks Formations, oil deposits that lie beneath parts of Montana, the Dakotas, and extending up into southern Saskatchewan. The good news is that the quantity of petroleum in these two geologic formations is, in the words of the U.S. Geological Survey, "larger than all other current USGS oil assessments of the lower 48 states" and "the largest 'continuous' oil accumulation ever assessed by the USGS." The good news becomes even better in light of a 2013 reassessment of reserves potential. USGS analysts reported that the formations contain an estimated mean of 7.4 billion barrels of undiscovered, technically-recoverable oil, an estimate 50-fold greater than had been reported 18 years earlier. The bad news in all of this is that 7.4 billion barrels are roughly equivalent to only a 13-month petroleum supply for the United States based on current levels of consumption.

Globally, supplies of petroleum and biofuels are expected to be adequate to meet the world's demand for liquid fuels for at least the next 25 years. Beyond that point, there is considerable uncertainty regarding consumption and available supplies. Natural gas supplies, on the other hand, are quite abundant with current reserves equivalent to more than several hundred years of supply at current levels of extraction.

In a 2016 report, the International Energy Agency forecast significant shifts in fossil fuel use, with the portion of global energy supplied by petroleum predicted to fall to levels of the late 1990s, and coal to mid-1980s levels. Natural gas is the only fossil fuel expected to increase as a percentage of global energy consumption. The effect of Trump administration policies in the U.S. with regard to coal could impact this outlook.

All projections are based on the assumption that fossil fuels will be fully available throughout forecast periods. How this plays out long term depends upon the climate change issue and society's response to it.

Petroleum Import Dependence

The U.S. domestic energy outlook for the decades ahead is significantly more positive than only several years ago. From a net importer of 60 percent of petroleum supplies and 30 percent of overall energy requirements in 2005, the United States is on track to reduce its energy import dependency to less than 10 percent by 2040, and to perhaps become a net exporter of energy resources thereafter. In 2015, the U.S. was a net importer of 24 percent of petroleum needs and 9 percent of total energy supplies.

Net Oil and Natural Gas Import Dependency of Selected Nations, 2013 and 2040

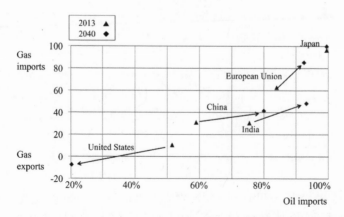

© International Energy Agency, *2015 World Energy Outlook,* IEA Publishing
(License: www.iea.org/t&c/termsandconditions).

While new technologies for obtaining oil and natural gas are available worldwide, deposits of energy resources are uneven, with two of the world's largest consuming regions—Asia and Western Europe—having relatively little in the way of reserves. As a result, in contrast to the United States, the import dependence of most major oil and natural gas consuming nations is likely to increase in the decades. Asia will emerge as the primary destination for oil and gas exports, and also account for the majority of growth in consumption.

Renewable Energy

Observers tend to discount the potential of wind, solar, and geothermal because together they still account for less than 2.5 percent of U.S. energy consumption after decades of investment, research, and technology development. However, a 2012 assessment by the National Renewable Energy Laboratory reveals vast renewable energy potential.

U.S. Estimated Technical Potential Generation

TECHNOLOGY	GENERATION POTENTIAL (TWH)
Urban utility-scale photovoltaic (solar)	2,200
Rural utility-scale photovoltaic (solar)	280,600
Rooftop photovoltaic (solar)	800
Concentrating solar power	116,100
Onshore wind power	32,700
Offshore wind power	17,000
Enhanced geothermal systems	31,300

As a basis for comparison, total electricity sales in the U.S. in 2010 were 3,754 TWh.

Source: Lopez et al., National Renewable Energy Laboratory (2012).

Estimated energy generation values are technology-specific which consider topographic, environmental, and land-use constraints, but not market or economic constraints, and thus likely overstate what may actually be possible. The numbers are nevertheless impressive, indicating domestic electricity generating potential from renewable energy more than 125 times greater than 2010 U.S. electricity consumption. The implications for the U.S., and the world, are large.

Viewed as having the greatest potential with respect to solar energy are 1) a network of rural solar power plants, and 2) large-scale solar generating facilities concentrated in desert regions where cloudy days are rare. Wind and what are described as enhanced geothermal systems also have technical potential well beyond current levels of use. Important to further development of renewable electricity are breakthrough developments in storage capability that would allow uninterrupted power supply over extended periods of calm winds and dark overcast days. Significant initiatives throughout the world are focused on this issue.

In addition to wind, solar, and geothermal, many see additional potential in bioenergy. For instance, a 2016 assessment of sustainable biomass energy potential in the U.S. indicated the possibility of replacing up to 30 percent of domestic petroleum consumption by biofuels. As noted previously, whereas current production of biofuels, such as ethanol, is relatively inefficient, yields of what are known as second generation biofuels will be far greater, providing much more fuel (5 to 10 times more) per ton of biomass input. A recent global assessment suggests even greater promise for biomass-derived energy. This study concluded that 50 percent of the world's primary energy could be sustainably supplied by biomass energy, and that doing so would require use of only 6 percent of the annual global production of biomass.

Current technology provides a number of options for conversion of biomass and other biomaterials to energy. These include direct firing for electrical generation, production of a variety of liquid fuels including ethanol and bio-diesel, repackaging of biomass in the form of fuel pellets for use in home heating, and use as a fuel in steam generation for either large-scale district heating or for powering manufacturing operations. Over 25 million tons of fuel pellets were produced globally in 2014, mostly from wood. About 6 million tons were produced in North America, with most exported to Europe (primarily the UK) and to Asia (primarily Japan). Fuel pellet production has become controversial in both North America and Europe based on concerns about impacts on forests and long-term sustainability of rising pellet consumption. Recent analyses suggest considerable potential for increased production based on net forest growth trends and the likelihood of increased extent of forest cover (at least in the U.S. South) linked to market driven incentives for forest expansion.

Nuclear Energy

Because of well-known safety risks and disposal issues, nuclear energy is often dismissed when considering future sources of energy. Nonetheless, 438 commercial nuclear reactors were in operation worldwide in 2015 in 30 countries, providing about 11 percent of global electricity consumption. An additional 67 reactors were under construction. Smaller-scale reactors (over 200) are found in research facilities around the world, with as many as several hundred more used to power ships and submarines.

France is more reliant on nuclear energy than any other country, obtaining about three-quarters of its electrical power from this source. Other countries relying heavily on nuclear power are concentrated in Europe. Those producing a quarter or more of electricity via nuclear generation include Belgium, Bulgaria, the Czech Republic, Finland, Hungary, Slovakia, Sweden, Switzerland, Slovenia, Ukraine, and South Korea. About one-fifth of electricity production in the U.S., U.K., Russia, and Spain is nuclear. Germany, a country long among the top ten producers of nuclear energy, decided to phase out of nuclear power in the aftermath of Japan's Fukushima nuclear disaster of 2011; all reactors in Germany are scheduled to be shut down by 2022.

Energy and Environment

None of the current sources of energy are without environmental problems. Problems linked to fossil fuels are well known. Combustion of coal results in emissions of large quantities of carbon dioxide, sulfur dioxide, and nitrogen oxides, and also releases of mercury. There is also a significant coal ash disposal issue, with 140 million tons of toxic ash generated each year at coal-fired power plants in the U.S. alone. Emissions from burning of oil in power plants include the same compounds as emitted in coal combustion, although in lesser quantities per unit of energy produced. Natural gas combustion yields nitrogen oxides, carbon dioxide, and emissions of methane, a potent greenhouse gas, resulting from incomplete combustion or leaks in transportation.

Complete combustion of gasoline yields only water, nitrogen, and carbon dioxide. Surprising to many is how much carbon dioxide results from combustion of a single gallon of gasoline. A gallon of gasoline weighs only 6.3 pounds. But its combustion yields 19.6 pounds of carbon dioxide, the result of oxygen reacting with hydrocarbons. Similarly surprising is that an average vehicle in the U.S. releases about 4.7 tons of carbon dioxide each year. Moreover, if combustion is incomplete, as is often the case, then vehicle emissions include volatile organic compounds (VOCs), nitrogen oxides, and carbon monoxide.

Unfortunately, no method of energy production is completely free of environmental impacts. The least environmental concerns are linked to geothermal, hydroelectric, and solar energy. Electricity generation or heat capture from geothermal sources comes closest to being environmentally benign, yielding no emissions. Hydroelectricity production is also associated with very low impacts, although building of dams can be resource intensive and is often controversial because of impacts on local residents, flora, and fauna as new areas are flooded. Also relatively low impact is solar energy, except for large land areas needed for solar energy collection, bird fatalities when solar collectors are linked to collection towers, and impacts associated with building and maintaining the solar collectors. Wind energy is also known to kill birds. In addition, some find the wind towers aesthetically objectionable, and the noise and strobe-effects of the spinning blades irritating. Then there are the impacts of constructing and maintaining the turbines and supporting towers. A single wind tower contains about 71 tons of materials, including an astounding quantity of copper; some 5,600 to 14,900 pounds of copper are used per MW of electricity generated.

Biomass energy has gained in popularity in recent years due to renewability and relative carbon neutrality when the biomass is obtained from sustainably managed farms and forests. Primary impacts occur in the process of harvesting and in combustion if particulate emissions are not controlled.

Despite the substantial risks associated with plant operation and storage of spent fuels, a positive aspect of nuclear power is that electricity generation does not produce carbon emissions. In a carbon-constrained economy, which could result from a societal focus on climate change, nuclear energy may well play a larger role in the global energy picture than today.

Issues and Options

Global supplies of energy appear to be adequate for the next several decades despite rapid consumption growth, a situation that could change dramatically should climate issues move to the fore in society. Climate issues or not, sufficiency of transportation fuels pose a challenge for the not-too distant future. Production of liquid fuels from biomass, greater use of electric powered vehicles, and conversion of vehicle engines and fuel distribution networks to allow use of natural gas are likely to provide at least part of the solution. In addition, in transportation, industry, residential and nonresidential buildings, and all other areas of energy use, efforts to conserve energy will become more important.

Regarding climate, if future developments lead society to conclude that fossil fuel use must be curtailed, then implementation of some form of carbon tax or caps on carbon emissions is a likely development. The effect would be to increase the prices of fossil fuels, thereby increasing the economic viability of alternative forms of energy. The speed and extent to which a transition to alternative energy could realistically occur is an open question.

References

Abt, K., R. Abt, C. Galik, and K. Skog. 2014. *Effect of Policies on Pellet Production and Forests in the U.S. South.* USDA-Forest Service, General Technical Report GTR-202.

Association for the Study of Peak Oil and Gas (ASPO). 2014. "Top Ten Developments of 2010."

British Petroleum. 2015. *Statistical Review of World Energy 2015,* Accessed February 5, 2016.

Campbell, C. 2003. *Oil Depletion—the Heart of the Matter.* Association for the Study of Peak Oil and Gas.

Froggatt, A., and G. Lahn. 2010. *Sustainable Energy Security: Strategic Risks and Opportunities for Business.* Chatham House-Lloyds 360° Risk Insight White Paper, June.

Gold, R., and A. Davis. 2007. "Oil Officials See Limit Looming on Production." *Wall Street Journal,* November 19.

Hart, P. 2010. "Oil Demand in the West to Decline, According to the IEA." *Oil Drum,* November 18.

Hirsch, R., R. Bezdek, and R. Wendling. 2005. *Peaking of World Oil Production: Impacts, Mitigation, & Risk Management.* A special report commissioned by the U.S. Department of Energy.

Hubbert, M. 1949. "Energy from Fossil Fuels." *Science,* 109: 103–109.

Hughes, J. 2012. *The Energy Sustainability Dilemma: Powering the Future in a Finite World.* Cornell University, May 2. (Considerable data attributed to Dr. Arnulf Grubler)

International Energy Agency. 2007. "The Next 10 Years are Critical—the World Energy Outlook Makes the Case for Stepping up Co-operation with China and India to Address Global Energy Challenges," *IEA News Release,* November 7.

International Energy Agency. 2007. *World Energy Outlook.*

International Energy Agency. 2014. *Key World Energy Statistics.*

International Energy Agency. 2015. *World Energy Outlook.*

International Energy Agency. 2016. *World Energy Outlook.*

International Energy Agency. 2015. *Energy Poverty.*

Kerr, R 1998. "The Next Oil Crisis Looms Large- and Perhaps Close." *Science* 281: 1128–1131.

Ladani, S. and J. Vinterbäck. 2009. *Global Potential of Sustainable Biomass for Energy.* Swedish University of Agricultural Sciences, Department of Energy and Technology, Report 013.

Lopez, A., B. Roberts, D. Heimiller, N. Blair, and G. Porro. 2012. *U.S. Renewable Energy Technical Potentials: A GIS-Based Analysis.* National Renewable Energy Laboratory, Technical Report.

Maechling, C. 2000. "Pearl Harbor: The First Energy War." *History Today* 50(12):41–47.

National Petroleum Council. 2007. *Facing the Hard Truths about Energy: a Comprehensive View to 2030 of Global Oil and Natural Gas.*

Oil Drum. 2010. "World Oil Capacity to Peak in 2010 Says Petrobas CEO." *The Oil Drum,* February 5.

Pollastro, R., L. Roberts, T. Cook, and M. Lewan. 2008. *Assessment of Undiscovered Technically Recoverable Oil and Gas Resources of the Bakken Formation, Williston Basin, Montana and North Dakota, 2008.* U.S. Geological Survey Open-File Report 2008-1353.

Schultz, S. 2010. "Military Study Warns of a Potentially Drastic Oil Crisis." *Spiegel,* September 1.

Shell Oil Company. 2011. *Shell Energy Scenarios to 2050.*

Strong, Z. 2015. "Copper Use in Renewable Energy Expected to Increase Dramatically as U.S. Legislates Upgraded Energy Policies." *Mining and Power Magazine.*

U.S. Department of Energy. 2011. *U.S. Billion-Ton Update: Biomass Supply for a Bioenergy and Bioproducts Industry.* R.D. Perlack and B.J. Stokes (Lead authors), ORNL/TM-2011/224. Oak Ridge National Laboratory, Oak Ridge, TN.

U.S. Department of Energy. 2015. "Monthly Energy Review," Energy Information Administration, December.

U.S. Department of Energy. 2014a. *Annual Energy Outlook 2014.*

U.S. Department of Energy. 2014b. *International Energy Outlook 2014.*

U.S. Department of Energy. 2016. 2016 Billion Ton Report: Advancing Domestic Resources for a Thriving Bioeconomy. Volume 1: Economic Availability of Feedstocks. Langholtz, M., Stokes, B. and Elton, L. (leads), ORNL/TM-2016/160. Oak Ridge National Laboratory, Oak Ridge, TN.

U.S. Geological Survey. 2008. "3 to 4.2 Billion Barrels of Technically Recoverable Oil Assessed in North Dakota and Montana's Bakken Formation—25 Times More Than 1995 Estimate," USGS News Release, April 10.

U.S. Geological Survey. 2013. "USGS Releases New Oil and Gas Assessment for Bakken and Three Forks Formations—Finds Formations Have Greater Resource Potential than Previously Thought," Accessed May 10, 2015.

U.S. Joint Forces Command. 2010. *The Joint Operating Environment 2010.*

Wood, J. and G. Long. 2004. *Long Term World Oil Supply Scenarios.* US. Department of Energy, Energy Information Administration.

PART 4

AVOIDING RESPONSIBILITY

We have seen that a lack of understanding of global realities and trends, and even of the basics of compounding, contribute to the modern equivalence of Paraíso. Unfortunately, a number of other factors including hypocrisy, political expediency, and wishful thinking are significant contributors as well.

The two chapters that follow explore the ways in which people rationalize social behavior. Also explored is the record of political leaders—namely U.S. presidents—in seeking to balance competing viewpoints and to perhaps change the narrative.

CHAPTER 10

Seeking Silver Bullets

People are always looking for the single magic bullet that will totally change everything. There is no single magic bullet.

—TEMPLE GRANDIN

If we only had fewer, simpler, less entangled problems, perhaps we could find a silver bullet solution that could be quickly applied, requiring little or no effort on anyone's part. Such solutions are periodically proposed, with advocates sometimes suggesting wonderful, instant results and no downside risks. While it is comforting to imagine such solutions, the reality is that the problems facing society today tend to be complex and multifaceted, with solutions requiring new knowledge, technologies, approaches, and sometimes compromise among competing objectives.

A number of silver bullet solutions related to environmental problems have been floated in recent years. Three of these are the focus of this chapter:

- Tree-free paper
- Bamboo flooring and rapidly renewable materials
- Recycling

Tree-Free Paper

In some circles, there has been growing interest in moving away from reliance on trees and wood as a papermaking material and toward agricultural-based or other raw materials. The movement is personified by actor Woody Harrelson, who supports paper made from wheat straw, and the San Francisco-based Conservatree, that promotes paper products made of anything other than wood fiber. The primary driving force behind what is widely referred to as "tree-free paper" campaigns appears to be a desire to save forests and trees.

Would reducing or avoiding altogether the use of wood-derived fiber in making paper, in fact, lead to more extensive forests and more trees? While intuition might suggest that the answer is obvious and "yes," further examination reveals there are aspects of wood fiber production that require more than casual consideration. For instance, over 90 percent of virgin wood fiber used in papermaking in the United States is sustainably harvested from privately-owned forest land from which many derive a significant portion of their annual income through harvest revenue. With this in mind, suppose that large numbers of people across the United States opted to use plastic paper or paper made of wheat straw rather than paper made from wood fiber. All things considered, would the nation be likely to have more trees and forests, or less of both? The proposition is similar to one in which people might be encouraged to eat apples rather than oranges with a goal of "saving" oranges. A wholesale shift to apples, of course, would simply lead private landowners to remove orange trees and convert the land on which they stand to apple orchards. The reality with respect to trees, forests, and forest owners is quite similar. Faced with a loss of markets and income from sales of wood and forest products, many owners of private forest land could be expected to convert what are now forests to cropland and pastures, or to sell lands to developers, with fewer trees and a smaller forest estate the result. In other words, systematically avoiding the use of wood is likely to lead to forest loss rather than the other way around.

A recent example of reduced wood markets leading to loss of forests can be found in Minnesota, where several closures of large wood-using mills in recent years have resulted in the sale of large forest industry-owned blocks of timber. Some of this land has been purchased by an agribusiness company that has begun clearing large areas of forest for conversion to potato production. Other land is being purchased for cabin and homestead development. Similar pressures are a fact of life in all forested areas of the country. A 2012 assessment by the USDA-Forest Service indicated potential losses of 21 million acres of forestland in the Southeastern region by 2060 due to urbanization and conversion to agriculture.

Beyond the issue of whether going tree-free in paper purchasing will actually save trees, a more fundamental question is what the environmental impacts are of alternative fiber production systems. The fact is that production, procurement, transport, and processing of *all* raw materials results in environmental impacts. It is therefore important to identify, and to the extent possible quantify, likely impacts before making conclusions about which types of paper may be environmentally preferable to other types of paper. This same logic applies to evaluation of any raw material source or production process. This issue is examined further in the following section.

Comparison of Environmental Impacts

In addition to concerns about loss of forests, another reason sometimes given for favoring paper made of non-wood fiber is dislike for the impacts of harvesting trees. Pulpwood detractors often point out that pine plantations are often monocultures, have low biodiversity, require use of chemical inputs, and are typically clearcut at the time of harvest. Fiber sources promoted as alternatives to wood fiber in papermaking include dedicated fiber crops such as hemp, bamboo and kenaf, and agricultural crop residues such as wheat straw. Each of these will be briefly examined for the purpose of understanding environmental impacts through the fiber production cycle.

Production of Wood Fiber

Pulpwood for paper production is typically obtained by:

- Thinning forest stands that are being grown to provide raw material for multiple products including sawlogs.
- Clearcutting patches of smaller, faster growing tree species, and/or recovering residues (wood chips) from sawmills. After harvesting, forestlands are often replanted; but some species, such as aspen, regenerate from their own root systems while other species regenerate from natural seeding.

Two-thirds to three-fourths of pulpwood used in U.S. paper production is obtained from the Southeastern region of the U.S. The majority of fiber comes from several common species of pine, known collectively as the southern pines.

Thinning in southern pine stands being managed for multiple products, including sawtimber, typically occurs at 8–10 year intervals until the trees reach 20–30 years of age. Each removal, beginning with the second thinning treatment, results in an increasing proportion of wood used in producing lumber, plywood, and other long-lived products. A stand-clearing harvest (clearcutting) occurs at 30–40 years of age, or about a decade after the final thinning. This is followed by direct seeding or replanting. Stand reestablishment is typically rapid.

When replanting, it is common to plant at a relatively high density in order to shade out other plants that compete with the trees for water and nutrients. Herbicides are sometimes also used prior to or soon after replanting. Later, the trees are thinned periodically to manage competition for water, sunlight, and nutrients and maintain forest health.

14-Years after Clearcutting of Southern Yellow Pine in Louisiana Followed by Replanting

Photo by J. Bowyer

Fertilizer may also be applied once or twice, with the first application at or near the time of planting, and a second application at age 4–7 if necessary. Thereafter, stand entries occur in conjunction with thinning cycles. Overall, during a 30–40 year growth/harvest cycle of southern pine, site entry occurs six to ten times.

Pine managed only for pulpwood is grown on a harvest cycle of about 20 years. In this case, the number of stand entries during the growth/harvest cycle is three to eight. Typical southern pine yields through a full rotation are 2.5 to 3.8 dry tons of wood per acre per year, a number that has increased steadily over the years as a result of genetic improvement in planting stock and improved management practices.

Outside of the Southeast, which enjoys a more generous growing season and higher tree growth rates than other regions, pulpwood is commonly obtained from less intensively managed forests (i.e., with fewer stand entries). For example, in the Lake States region, fast-growing and short-lived aspen (typical life expectancy is 70–80 years) is harvested at 40–50 year intervals. In the case of typical aspen pulpwood management, the number of stand entries in the 40–50 year growth harvest cycle is one to two. Aspen is a pioneer species and requires direct sunlight early in the growth cycle. Harvest is, therefore, commonly done by patch clearcut with the desired size of the clearing being at least twice the height of the surrounding forest to ensure that the opening receives full sunlight. Because aspen re-sprouts from its root system (root suckering), no replanting or seeding is necessary. The new trees are generally about 5–6 feet in height one year after harvest. Herbicides and fertilizer are also generally unnecessary because new trees regrow fast enough and in sufficient density to outpace competing shrubs, plants and grass. Thinning is uncommon. Aspen typically yields 0.4 to 0.9 dry tons per acre per year at a harvest age of 40–50 years.

Clearcutting in Minnesota Aspen and Site Recovery without Replanting.

Strip Clearcutting in Aspen

Photo by J. Bowyer

Aspen Site One Year Following Clearcut Harvest

Photo by J. Bowyer

Although a number of studies have found that tree plantations, such as those from which pulpwood is obtained, often provide less suitable habitat for a number of plant and animal species than do natural forests, plantations are far from biological deserts. Research generally recognizes lower diversity of life in plantations, but also acknowledges that plantations can create critical habitat and wildlife corridors. Some refer to plantations as a "lesser-evil" alternative to urban development and agriculture, while stressing the importance of native rather than exotic tree species. A recent and frequently referenced evaluation of this issue concluded that "there is abundant evidence that plantation forests can provide valuable habitat, even for some threatened and endangered species, and may contribute to the conservation of biodiversity . . ." In fact, establishment of planted forests is commonly advocated as a way to reestablish wildlife in agriculturally dominated areas.

Agricultural Fiber Crops

The most widely promoted non-wood papermaking fibers are bamboo, kenaf and hemp. All of these are obtained from dedicated fiber crops.

The most extensively studied alternative fiber crop is kenaf, a fast-growing annual crop that can reach heights of 18 feet or more in a single season. An intensive USDA crops screening program involving more than 400 fibrous plants that were viewed as having potential to expand and diversify markets for American farmers, led to identification of six promising crop species: kenaf, crotalaria, okra, sesbania, sorghum, and bamboo. Kenaf (*Hibiscus cannabinus* L.) emerged as the top candidate for further research, and subsequent investigation led to extensive knowledge of technical and economic aspects of plant growth and harvest, storage, and conversion to pulp and paper products. While hemp was not investigated as part of this work, other studies have shown kenaf fiber to have greater utility as a papermaking material.

A problem with both kenaf and hemp is that the stalks from which fiber is obtained are not uniform in fiber properties. Both have outer layers that are made up of long fibers as long as or longer than wood fiber, but inner core sections that make up two-thirds of volume in which fibers

are exceedingly short. This is a particular problem in hemp. Still, both of
these crops and also bamboo can be used alone or as supplemental fiber
to wood in production of paper.

In view of the fact that part of the motivation for developing tree-free
paper is concern for the environment, it is worth considering what environmental impacts are linked to alternative fiber production. Advocates
of tree-free paper sometimes claim that yields of fiber crops are far
higher than forests or tree plantations, and that fertilizer, pesticides, and
herbicides are not necessary in their production. However, with a few
exceptions, field experience has shown that differences in annual fiber
yields are much smaller than often suggested, and that fertilizer/pesticide/
herbicide applications are common because they greatly increase
growth rates, yields, and economic returns. Moreover, because fiber
crops are planted and harvested annually, as opposed to tree plantations
that are planted only once each 20–40 years and tended only a few times
during the fiber cycle, the volume of inputs and frequency of site entry are
also greater. Consider, for instance, the steps taken annually to produce a
crop of kenaf:

- preparation for planting (chisel, disc, disc/herbicides/disc, done twice)
- application of pre-plant fertilizer
- bedding
- planting of seeds
- application of side-dressing
- cultivation
- harvesting

This sequence of production steps results in direct-site impacts, including soil disturbing activities, about 700 times over a 50-year period, and
includes annual applications of fertilizer and herbicides. This is similar to
the cropping sequence involved with production of hemp. Production of
bamboo involves slightly fewer inputs, but in any case would likely involve 200–300 passes across the landscape over a 50-year period. In contrast, production of wood in natural forests or forest plantations typically

involves fewer than 10 site entries or soil disturbing activities over a 50-year period.

All of these fiber crops—kenaf, hemp, and bamboo—are typically planted as monocultures in vast plantations that are annually removed from the land at the point of harvest (i.e., by clearcutting). Plant and animal diversity is extremely low.

Agricultural Crop Residues

Crop residues have been used in papermaking for many years across the world in countries such as China, India, Pakistan, Mexico, Brazil, and more. U.S. research examining potential uses of crop residues as a papermaking raw material dates back to at least World War II. In the 1940s, 25 mills in the Midwest produced almost one million tons of corrugating medium annually from straw. Momentum in the U.S. non-wood fiber industry was lost following the war because of the high costs of gathering and storing straw and other issues. More recent research has shown that cereal straws, such as wheat, barley, and oats are the most promising source of crop residue fiber. These agricultural feedstocks have been found to be best suited for short-fiber applications as they behave similarly to recycled fibers.

In 2016, the United States produced 67.2 million metric tons of wheat, barley, and oats. About 92 percent of this production was accounted for by wheat. For a number of reasons, only a small quantity of crop residues is potentially available for new uses. Typically about 60 percent of straw produced is either used to feed livestock or is left on the field for soil conservation purposes, leaving 40 percent available for other uses. Variability in weather can markedly change straw availability from year to year. In persistently dry regions, soil conservation concerns may dictate no straw harvest. Even considering these caveats, there is a potential significant volume of available straw. For example, assuming a straw yield of 15 percent from all wheat growing areas in the U.S. equates to an estimated 10.1 million metric tons of annual straw yield. This is equivalent to about 15 percent of U.S. pulpwood production in 2012.

While the same issues that led to cessation of the use of straw in papermaking in 1940 remain issues today, it would nonetheless be possible to manufacture a certain percentage of paper from crop residues, or at least to use mixtures of wheat straw and wood pulp in producing paper. Even though there are few concerns from an environmental perspective in producing paper from crop residues rather than from wood, one additional factor that must be considered in evaluation of environmental impact is crop yields. Given that there are environmental impacts associated with all raw materials, those impacts need to be weighed against the quantity of the material produced.

Intuition might suggest that annually harvested crops should yield far greater quantities of fiber over a period of time than trees that are harvested only occasionally. In this case, intuition is wrong. When cumulative volumes of fiber produced from multiple years of annual harvests of fiber crops are compared with the volume of wood fiber produced in the course of several thinnings and a stand-clearing harvest at 20–40 years, total fiber yields from fiber crops and tree plantations are generally quite similar. Both fiber crops and tree plantations produce greater fiber yields than natural forests.

Recorded yields for annually harvested hemp, for instance, are only one-half to two-thirds those of plantation southern pine harvested on a 20-year cycle, while bamboo yields are comparable. In other words, a single harvest of pine provides about the same fiber yield as 20 harvests of bamboo. Kenaf yields range from 30–40 percent greater, to near the same, to as little as one-half that of plantation pine. Fiber crop yields are in all cases better than those from natural forests.

Is tree-free paper environmentally better than paper made from fiber obtained from forests and tree plantations? When all factors are considered, what may have intuitively appeared obvious is not so clear. In fact, a strong argument can be made that producing paper from trees makes the most sense from an environmental point of view.

Bamboo Flooring and Rapidly Renewable Materials

Produced in limited volume in China beginning around 1990, it was more than a decade later that bamboo flooring began to make inroads into world

floor covering markets. Marketing efforts centered on claims of comparable properties and environmental superiority in comparison to wood
flooring. A significant turning point occurred in November 2002 when
the U.S. Green Building Council designated rapidly renewable materials,
including bamboo, as *environmentally preferable materials.* This designation appeared in its Leadership in Energy and Environmental Design
(LEED) v.2.1 green building standard. Within three years bamboo flooring production had increased significantly in China, foreshadowing a
much larger production expansion. The "green" bamboo genie was out
of the bottle.

About two thirds of bamboo flooring in 2005 was produced in China,
with the vast majority going to the U.S. and the European Union. In the
U.S., the number of suppliers of bamboo flooring rose from fewer than 10
in the late 1990s to about 200 by 2005, with imports in 2005 approximating 45 million square feet. Similar growth in bamboo flooring consumption occurred in the EU, although total sales amounted to only one-fifth
to one-sixth that of the U.S.

Anointment of bamboo as a green material was based on broad and un-
questioned acceptance of the idea that rapidly-grown materials (i.e., those
that renew in 10 years or less) are somehow inherently environmentally
superior to those that renew in 11 years or more. It was a concept that was
embraced by LEED through the 2009 version of its standard, as well as a
number of other green building programs patterned after the LEED program. These programs assign green points for use of non-certified bamboo
along with environmentally-certified wood in construction. Even governments got into the act. For instance, the Governmental Construction
Organization of the Netherlands (*Rijks Gebouwen Dienst*) also accepts
non-certified bamboo as a green material.

In 2005, as now, the websites and promotional literature of bamboo
promoters were rife with glowing claims about the environmental attributes of bamboo. Currently, environmental claims are a bit more re-
strained than in 2005, but misinformation continues to be disseminated
via a number of sources. For example, an often-cited 2008 article that
appeared in *Scientific American* states:

Bamboo's environmental benefits arise largely out of its ability to grow quickly—in some cases three to four feet per day—without the need for fertilizers, pesticides, or much water . . . Bamboo is so fast growing that it can yield 20x more timber than trees on the same area.

These same claims can be found on a number of promotional websites. One site references the "vast supply" of bamboo, stating ". . . there are reportedly more than 1.6 million square miles of bamboo growing in China alone, with most of these native forests owned and managed by the Chinese government." With regard to this latter statement, China's bamboo resource in 2010 was reported by the UN Food and Agriculture Organization (FAO) to be a little over 14.1 million acres, an area that is equivalent to slightly over 22,000 square miles, the vast majority of which is plantations that are managed by farmers. Rapid renewability is a recurring theme in these and other websites.

From a social perspective, expansion of bamboo resources and industrial production has been quite successful. Numerous studies have documented expansion of household incomes and poverty reduction through increased bamboo production, and planting of bamboo on steep slopes previously terraced for agricultural production has helped to stabilize these slopes and reduced runoff and erosion in many areas. These benefits have been aided by development of standards for reforestation.

The bamboo phenomenon, however, has been accompanied by substantial environmental costs. Summarized below are observations of a number of investigators and research teams.

- Impacts on Biodiversity. Many observers have noted that bamboo forests in many parts of southwest China have become monoculture plantations, with the accompanying negative effects on local ecosystems. This has come about because bamboo growers are increasing the density of bamboo culms per acre, which effectively creates monocultures. The end result may be that, in the long run, the forests will be less resistant to threats, such as pests, disease and weather. There are also concerns about possible reductions in erosion control and nutrient cycling, which will in turn potentially lower productivity of bamboo forests. In particular, one species, Moso bamboo, has become dominant, and is threatening other species in some areas.

- Impacts on Forests and Farmland. Because bamboo is profitable to farmers, some have begun converting farmland back to bamboo forest to the extent that China's food security could be at risk. In addition, some mountainous forest areas have been clearcut to grow bamboo without consideration of site condition and future market changes.

- Water and Soil Concerns. Some have expressed concerns that bamboo may, in some areas, be an inappropriate forest type, which could lead to heavy soil erosion and transport of nitrogen and potassium into nearby rivers and water systems. Others note that the lack of biodiversity may decrease soil microbial activity, citing that in some Moso forests, natural soil fertility and site quality have gradually declined. Water retention issues have also been noted.

- Use of Fertilizer, Pesticides and Intensive Management. To achieve the desired density of bamboo culms, the land must be cleared of other species. This requires two brush cuttings each year in addition to topsoil tillage every one or two years and application of chemical fertilizers and pesticides in varying quantities. Bamboo, because it is a fast-growing plant, requires large quantities of nutrients, such that, on average, farmers are applying 178 pounds of fertilizer per acre (200 kg per hectare) each year.

Bamboo is a marvelous resource that provides multiple benefits for billions of people. Development of bamboo resources is economically assisting impoverished people while at the same time stabilizing erodible slopes and flood-prone watersheds. The ability to significantly accelerate growth through intensive management for commercialization purposes magnifies its many benefits. The benefits, however, come at a high environmental cost. Degradation of natural forests, tremendous biodiversity loss, widespread use of fertilizers and pesticides, loss of resilience in bamboo resources, and increased social and environmental risks linked to large-scale monoculture agriculture are among the costs.

The rapid renewal capacity of bamboo is a reality. But neither bamboo nor rapid renewability represent any kind of silver bullet. Equating rapid growth to environmental superiority without systematic consideration of measures employed to achieve rapid growth provides an example of the risks of casual forumulation of recommended practices.

Recycling

There is much to be gained from increasing recycling activity, and nothing that follows is intended to suggest that recycling is unimportant. However, recycling becomes problematic when it is suggested as *the* solution to the kinds of problems discussed herein. While quite important, carrying a few more bottles and cans to the curbside is only part of a much larger issue that centers on consumption.

In September 2014, Coca Cola Enterprises and Tesco, a global U.K.-based grocery retailer, teamed up to sponsor a campaign that was unfortunately dubbed ***Recycling is the Answer***. With the laudable goals of increasing the recycling rate in the U.K. and to reduce Coca Cola's carbon footprint, while also helping customers to do the same in their homes, the effort was focused on informing grocery shoppers and gaining commitments to change recycling behavior. In the words of Nick Brown, associate director of recycling for Coca-Cola Enterprises ". . . Past initiatives have demonstrated the power of pledging combined with awareness-raising, and we hope Recycling is the Answer will have a similar impact."

Recycling is certainly a significant challenge for the world's largest beverage distributor. With a reported 1.9 billion servings of Coca Cola products each *day* globally, the number of drink containers entering the waste stream as a result of this company's operations is simply staggering. In the U.K. alone, Coca Cola produces 2.58 *billion* cans, 450 million plastic bottles, and 67 million glass coke bottles of soft drinks and other beverages per year. The good news is that the recycled percentages of aluminum in cans and plastic and glass in bottles is 50, 25, and 37 percent, respectively. The good news gets even better in that the Coca Cola company is continually working to increase the recycled content of its aluminum drink containers and to increase the bio-based content of its plastic containers. The bad news is that currently 50, 75, and 63 percent of the material used each day in making new aluminum, plastic, and glass containers, respectively, is virgin material. The bad news gets worse in that sales of packaged drinks are growing very rapidly, having more than doubled worldwide in just the past two decades.

François Grosse addressed the promise of recycling and the realities of progress when consumption is rapidly increasing. He began by posing a question:

> *Could recycling play a major role in this decoupling between economic development and the need for material resources? Intuitively, the answer seems to be "yes": if we recycle massively, we will be reducing the consumption of virgin raw materials just as massively. For example, recycling 80% of a raw material means that the need for natural resource is divided by five. This impressive figure suggests that if a resource is recycled efficiently, it would take five times longer to exhaust this resource. A lifespan of 100 years would become 500 years thanks to recycling. The magnitude is certainly disputable in geological terms or in regard to the history of mankind, but it is still worthy of notice and esteem in terms of public policies and quite sufficient to serve as a springboard for energetic action. The very long-term problem is not entirely solved, but it does seem to be significantly lessened . . . Unfortunately, this rationale is faulty.*

He noted that when consumption is growing, for instance at a constant rate, then the shape of the consumption curve of a natural resource remains identical with or without recycling, with the only difference being that the curve is shifted to the right (i.e., that the magnitude of resource consumption is simply delayed). To make other than illusory progress through recycling then, the rate of increase in recycling needs to be greater than the rate of increase in consumption. Bottled water recycling statistics for the United States illustrate dramatically the result when the reverse is true—when the rate of consumption is rising more rapidly than the rate of recovery for recycling.

Effect of Recycling if the Annual Growth Rate
of Raw Material Consumption is Constant.

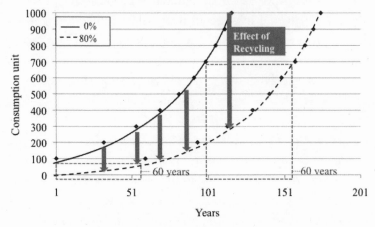

The annual consumption curve of virgin material is flattened by the effect of recycling. But when, after 60 years, consumption with recycling also overtakes the 100 value point, the dashed curve becomes identical to the other curve, except for a shift of 60 years. In this example, after 100 years, only 200 will be drawn from natural resources with recycling at 80 percent, whereas it would already be 700 without recycling. But the 700 mark will be reached only 60 years later with recycling and we will never gain more than 60 years with recycling if the progression of total consumption remains unchanged.

Source: Grosse (2010)

Between 1991 and 2010, recycling of polyethylene terephthalate (PET) plastic bottles used in packaging water increased in volume from 152,000 tons to 778,500 tons, a 5.1-fold (or 410 percent) increase. This is clearly a significant accomplishment and seemingly good news. But over that same time period, the recycling rate remained largely flat, increasing by a little over 5 percentage points overall (from 24.9 percent to 29.1 percent). The result was that the quantity of bottles *not* recovered (i.e., wasted) increased from 457,000 tons to 1.9 million tons. What at first glance appeared to be significant progress hardly looks like progress at all when the effects of increasing consumption are taken into account. A similar, though less dramatic, situation exists with respect to aluminum cans. In this case the recycling rate is about 50 percent, with this percentage having remained largely unchanged for a number of years.

Possibly accentuating this problem is what may be a tendency for people to consume more when there is a recycling option. Collaborative research between the Washington State University College of Business and the Tsinghua University Department of Industrial Engineering identified the phenomenon in two simple experiments. In one, people were given a pair of scissors to test and paper on which to test them. Those given the option to recycle the paper at the end of the test used more paper than those who were not. In the second experiment use of paper towels in public restrooms was found to be higher with the introduction of a recycling bin compared to when a recycling option was not provided. The researchers observed that:

> . . . *consumers may focus only on the positive aspects of recycling and see it as a means to assuage negative emotions such as guilt that may be associated with wasting resources and/or as a way to justify increased consumption. Therefore, an important issue would be to identify ways to nudge consumers toward recycling while also making them aware that recycling is not a perfect solution and that reducing overall consumption is desirable as well.*

Silver Bullets and a Better Future

Progress toward a sustainable future is not abetted by grasping at quick-fix solutions proposed by ardent advocates, especially absent thoughtful, systematic assessment of what is being proposed. There have always been those who have sought people's attention, money, or endorsement by offering miracle fixes, whether in the form of magic elixirs or new products with miraculous properties. The admonition "buyer-beware" is as applicable to those offering miracle environmental solutions as to those offering miracle cures.

References

Abrahamson, L. and L. Wright. 2000. "The Status of Short-Rotation Woody Crops in the U.S." In *Proceedings, XXI World Forestry Congress, Volume 2- Abstracts of Group Discussions,* edited by R. Jandl, M. Devall,

M. Khorchidi, E. Schimpf, G. Wolfrum, and B. Krishnapillay, p. 9. Rome: Food and Agricultural Organization of the United Nations.

Atchison, J. 1996. "Twenty-five Years of Global Progress in Non-Wood Plant Fiber Repulping." *Tappi* 79(10): 87–95.

Bambooki.com. 2011. "Is Bamboo a Green Material?" Accessed August 14, 2015.

Bowyer, J. 1999. "Economic and Environmental Comparisons of Kenaf Growth versus Plantation Grown Softwood and Hardwood for Pulp and Paper." In *Kenaf Properties, Processing and Products,* edited by T. Sellers, Jr., N. Reichert, E. Columbus, M. Fuller, and K. Williams, 323–346. Starkville, Mississippi: Mississippi State University.

Bowyer, J. 2001. *Industrial Hemp (Cannabis sativa L.) as a Papermaking Raw Material in Minnesota: Technical, Economic, and Environmental Considerations.* Department of Wood and Paper Science Report Series—2001, 50pp.

Bowyer, J., J. Howe, K. Fernholz, and P. Guillery. 2004. *Tree-Free Paper, When Is It Good for the Environment?* Dovetail Partners, Inc., Sept. 2.

Bowyer, J., K. Fernholz, M. Frank, J. Howe, S. Bratkovich, and E. Pepke. 2014. Bamboo Products and their Environmental Impacts: Revisited. Dovetail Partners, Inc., March 10.

Bremer, L., and K. Farley. 2010. "Does Plantation Forestry Restore Biodiversity or Create Green Deserts? A Synthesis of the Effects of Land-Use Transitions on Plant Species Richness." *Biodiversity Conservation* 19: 3893–3915.

Breyer, M. 2013. "Woody Harrelson Fights for the Forests with his Tree-Free Paper Company." Mother Nature Network, Accessed August 14, 2015.

Brockerhoff, E., H. Jactel, J. Parrotta, C. Quine, and J. Sayer. 2008. "Plantation forests and biodiversity: oxymoron or opportunity?" *Biodiversity Conservation* 17: 925–951.

Buckingham, K., L. Wu, and Y. Lou. 2013. "Can't See the (Bamboo) Forest for the Trees: Examining Bamboo's Fit Within International Forestry Institutions."Ambio Online, Accessed August 14, 2015.

Catlin, J., and Y. Wang. 2013. "Recycling gone bad: When the option to recycle increases resource consumption." *Journal of Consumer Psychology* 23(1): 122–127.

Container Recycling Institute. 2015. *PET Bottle Sales and Wasting in the US.*

Earth Talk. 2008. "Is Bamboo Flooring Better for the Planet than Traditional Hardwood?" *Scientific American,* Dec. 16.

Gleason, G. 2013. "The World of Tree Free Paper." Conservatree, Accessed August 15, 2015.

Gaiam. 2011. "How Eco-Friendly is Bamboo?" Gaiam Life, Accessed August 14, 2015.

Gallagher, S. 2011. "China's Appetite for Bamboo is Damaging Forests." Pulitzer Center China. Pulitzer Center on Crisis Reporting, Accessed August 15, 2015.

Grosse, F. 2010. "Is Recycling 'Part of the Solution'? The Role of Recycling in an Expanding Society and a World of Finite Resources." *SAPIENS Perspectives* 3 (1).

JiangHua, X., and Y. QingPing. 2012. *Shoot Plantation.* Research Institute of Subtropical Forestry/INBAR.

Kugler, D. 1990. "Non-Wood Fiber Crops: Commercialization of Kenaf for Newsprint." In *Advances in New Crops,* edited by Janick, J. and Simon, J., 289–292. Portland: Timber Press, Inc.

Malin, N., and J. Boehland. 2006. "Bamboo in Construction: Is the Grass Always Greener?" *AIA Architect.* April.

Maoyi, F., and Y. Xiaosheng. 2004. "Moso Bamboo (*Phyllostachys heterocycla var pubsecens*) Production and Markets in Anji County, China." In *Forest*

Products Livelihoods and Conservation: Case Studies of Non-Timber Forest Products Systems, Vol. I, edited by K. Kusters and B. Belcher, 241–158. Bogor, Indonesia: Center for International Forestry Research.

Mertens, B., L. Hua, B. Belcher, M. Ruiz-Perez, F. Maoyi, and Y. Xiaosheng. 2007. "Spatial Patterns and Processes of Bamboo Expansion in Southern China." *Applied Geography* 28(1): 16–31.

Metafore. 2006. *The Fiber Supply Technical Document, Summary Report.*

Morris, M. 2012. "Rapidly Renewable." *Ecobuilding Pulse,* July 19.

Panda Standard Association. 2012. *Forestation of Degraded Land Using Species Including Bamboo.*

Scott, A., Jr. and C. Taylor. 1990. "Economics of Kenaf Production in the Lower Rio Grande Valley of Texas" In *Advances in New Crops,* edited by J. Janick and J. Simon, 292–297, Portland, OR: Timber Press, Inc.

Siddiqui, K. 1994. "Cultivation of bamboos in Pakistan." *Pakistan Journal of Forestry* 44: 40–53.

Song, X., G. Zhou, H. Jiang, S. Yu, J. Fu, W. Li, W. Wang, Z. Ma, and C. Peng. 2011. "Carbon Sequestration by Chinese Bamboo Forests and their Ecological Benefits: Assessment of Potential, Problems, and Future Challenges." *Environmental Review* 19: 418–428.

Spence, Y. 2013. "Benefits of Bamboo." *HubPages,* Feb. 24, Accessed August 15, 2015.

van der Lugt, P., and M. Lobovikov. 2008. "Markets for Bamboo Products in the West." *Bois Forêts Des Tropiques 295.*

Yiping, L., and G. Henley. 2010. *Biodiversity in Bamboo Forests: A Policy Perspective for Long Term Sustainability.* International Network for Bamboo and Rattan (INBAR), Working Paper 59.

CHAPTER 11

An Environmental President

President Clinton will go down in history as one of the
great defenders of the environment.

—CARL POPE, PRESIDENT, SIERRA CLUB

A president of the United States has innumerable issues to deal with, myriad constituencies to consider. But one area, more than most, appears critical to establishing a positive and enduring legacy—the environment and environmental protection. Actions vis-à-vis the environment on the part of several U.S. presidents are briefly examined in this chapter. The actions of one, hailed by several major environmental organizations and news media as perhaps the greatest environmental president of all time, are explored in greater detail.

An Environmental Legacy in the Making

Just after noon on September 18, 1996, less than two months from the election that would determine whether he would serve a second term, President William J. Clinton rose and strode confidently to the podium. As a backdrop he had selected the Grand Canyon, and on this waning day of summer, the sky was sunny, the breeze warm.

A robust economy and strong job growth over the preceding several years had helped to place Clinton in a strong position with the voting public, and polls showed the President to be comfortably ahead of his Republican challenger in the upcoming election. However, many in his administration found it troubling that the President's environmental record had been increasingly and bitterly criticized during the re-election campaign. Criticism regarding Clinton's environmental record persisted despite actions in his first year of office to drastically reduce harvesting in the national forests of the Pacific Northwest, and later in national forests nationwide, and a presidential ban on road-building and logging in vast areas of the West.

Whether motivated by politics or a genuine desire to protect the environment or both, the President had come to El Tovar Lodge on the south rim of the Canyon to announce the designation of the 1.9 million acre Grand Staircase-Escalante region in Utah as the nation's newest National Monument. He would use the Antiquities Act of 1906 as the authority upon which to base the designation, citing in his remarks the fact that Theodore (Teddy) Roosevelt had used the same act in 1908 to provide long-term federal protection to the Grand Canyon.

In his announcement, Mr. Clinton talked about the natural beauty of the Escalante Canyons and the surrounding region and he noted that his designation would assure access of current and future generations of Americans to hike, camp, fish, and hunt. He also expressed concern about proposed coal mining development—the area is estimated to hold some 62 billion tons of low sulfur coal of which at least 11.4 billion are estimated to be economically recoverable using current technology, a recoverable quantity sufficient to supply the entire United States for over 11 years, and the ten westernmost states for over 93 years at 1996 rates of use. This factor was apparently the overriding reason for the monument designation. Clinton noted that Andalex Resources, a company that was engaged in environmental analyses required in the permitting process for coal mining, had agreed to trade its lease to mine coal in the Grand Staircase-Escalante region for "better, more appropriate sites" outside the boundaries of the newly designated monument.

What the President appeared to ignore or dismiss on that glorious September day is that the Grand Staircase/Escalante region is richly endowed with not only coal deposits, but a number of non-fuel minerals as well, including uranium, zirconium, titanium, and vanadium. Similarly, there was no mention in the President's speech of the large and growing U.S. dependence on imported energy and mineral resources.

Clinton's 1996 re-election set the stage for bold new environmental initiatives. With the encouragement of many within his administration, including Vice-President Al Gore, Interior Secretary Bruce Babbitt, and White House Chief of Staff John Podesta, the President sought to establish a legacy as an "environmental president." Clinton used the Antiquities Act 21 times during his second term in designating or expanding national monuments. Nine of these occurred in his final week of office. Like Grand Staircase/Escalante, several of these designations were specifically designed to prevent minerals development.

The President's actions appeared to resonate with the media and the environmental community. Shortly after Clinton left office, Carl Pope, executive director of the Sierra Club, confidently predicted that "President Clinton will go down in history as one of the great defenders of the environment," a sentiment echoed by the *Washington Post*'s William Booth who wrote "Clinton leaves office with what may be the most substantive environmental legacy of any president since Theodore Roosevelt." The Wilderness Society had earlier described Clinton "as one of the top conservation presidents of all time." Similar views were expressed in the *Seattle Post-Intelligencer,* the *Minneapolis StarTribune,* and other news media.

Strong words. Ringing endorsements. An enduring legacy. An environmental president—indeed *the* environmental president for the ages. If it were only that simple. Before printing the next generation of history books we should reconsider Mr. Clinton's actions vis-à-vis the environment. Three actions of the Clinton/Gore administration provide informative examples of the nature of thinking behind environmentally-related decision making that occurred over that eight-year period: 1) the spotted owl recovery plan, 2) designation of the Grand Staircase-Escalante National Monument, and 3) extending roadless designation to over 58 million acres of forest land in the western U.S.

The Spotted Owl Recovery Plan

In Mr. Clinton's inaugural year (1993) a cauldron of controversy swirled around the forests of the Pacific Northwest. Scientists had earlier recommended substantial harvest reductions, especially of older, larger trees, for the purpose of protecting populations of the northern spotted owl and the marbled murrelet. Interest groups representing every possible viewpoint vigorously agreed and disagreed, and the courts had intervened several times in the previous couple of years. The response of the Clinton administration was to sponsor a highly visible conference of selected stakeholders, followed by formation of a select team of scientists—overwhelmingly biologists—who were asked to develop alternatives for action. The President ultimately selected an option that called for reduction of timber harvests on federal lands of the Pacific Northwest from about 5 billion board feet annually to 1 billion. Yearly harvests subsequently declined to less than 0.3 billion, resulting in a more than 90 percent reduction in timber harvests from federal ownerships in Washington, Oregon, and northern California. The decision came to be known as the "Clinton Forest Plan."

At first glance, the plan that was adopted would appear to be a bold initiative that for once placed the environment ahead of the economy—clearly a milestone in environmental protection. Peter Berle, who in 1994 was president of the National Audubon Society, thought so. He described the forest plan as a model for environmental planning in the U.S. and the world. But how sound was the planning process and the principles upon which it was based? Should future environmental planning be modeled after this effort? And just how wise was the decision to massively reduce harvests?

In the discussions leading up to the decision to drastically reduce timber harvests on public lands in the Pacific Northwest, there was no consideration of consumption in the Pacific Northwest or other parts of the United States or of the supply of raw materials required to supply that consumption. In fact, the issue of consumption was not even on the table in any of the deliberations. As a result, a number of decisions of critical importance to the environment of the U.S. and the world were made without the benefit of any discussion at all.

When President Clinton decided to significantly reduce production of wood raw materials without a plan either to obtain replacement materials or to reduce consumption, a decision was made, by default, to obtain replacement raw materials somewhere other than the forests of the Pacific Northwest. And, because the U.S. was (and is) a net importer of all categories of materials—metals, cement (the basis for concrete), petroleum (the basis for most plastics), and wood and wood fiber, a decision was also made to increase our level of imports, again with really no discussion. There is no evidence that the President or anyone within his administration asked or considered any of the following questions:

- If harvest levels within the national forests of the Pacific-Northwest were reduced by four billion board feet annually, where might replacement wood come from?
- Is there any plan in place for reducing U.S. consumption of wood by four billion board feet annually?
- What would be the environmental impact if consumers switched from use of wood to use of steel, aluminum, concrete, and/or plastic?
- What are the environmental implications of producing four billion board feet annually in the likely new producing region?
- Are there rare or endangered species in the likely new producing region and will they be negatively impacted by increased harvest activity?
- What are the likely environmental impacts of shipping four billion board feet annually from the new producing region to regional and national markets?
- What are the risks of importing exotic pests as a result of substantial increases in softwood imports?

Although then Oregon Senator Patty Murray, citing "environmental imperialism," called attention to such issues in an impassioned speech on the Senate floor, these questions were neither addressed nor even on the table for discussion by the planning team. How is it possible that what has been described as the most comprehensive environmental planning process in history, led by an individual described as one of the top conservation presidents of all time, failed to consider such questions?

That these things were not considered or discussed does not change the reality that in adopting the Clinton Forest Plan, a decision was effectively made to trigger a number of substantial and negative environmental impacts both within and outside the United States. Within only several years of spotted owl plan implementation, timber production from the nation's national forests declined precipitously, net softwood imports into the U.S. increased sharply—an entirely predictable result that led, in turn, to a sharp rise in the U.S. wood and forest products trade balance deficit. In addition, softwood harvests in the U.S. Southeast exceeded growth for the first time in over six decades, steel gained 4 to 6 percent of the U.S. house framing market (up from 0 in 1990)—a development that significantly increased the quantity of energy needed to build a home, and the U.S. became a net importer of non-fuel mineral resources for the first time in modern history. Although a direct link between these developments and the spotted owl recovery plan cannot be proven, the initiative likely played a role in most or all of them.

Timber Harvest from U.S. National Forests 1985–2014
(Billions of board feet)

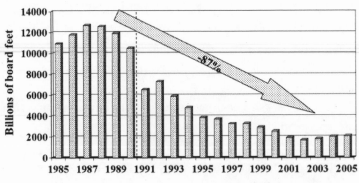

Source: USDA-Forest Service (2014)

U.S. Net Imports of Softwood Lumber 1985–2005

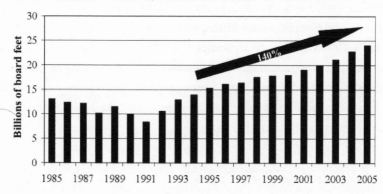

Source: Howard, J. U.S. Forest Products Laboratory (2003)
data for 2003–2005 from Random Lengths.

U.S. Trade Balance, Timber Products 1965–2004

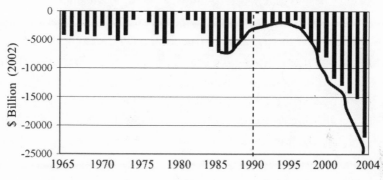

Source: Howard and Jones (2016)

Overall, during the Clinton administration, timber harvests from federal lands fell from 2 billion to 340 million cubic feet (an 83 percent decline), and the proportion of U.S. softwood lumber consumption provided by national forests fell from 24 percent to 3 percent. It was a stunning development. Coincidentally, U.S. net imports of timber and wood products increased almost 2.1 billion cubic feet, and softwood lumber imports more than doubled, during the same time period.

Subsequent to actions to protect habitat by sharply curtailing timber harvests, it was discovered that a larger, more aggressive species of owl—the barred owl—which had been expanding its range westward for decades, was responsible for significant reduction of spotted owl populations. An assessment published in late 2015 identified the spread of the barred owl as the primary reason for continued decline of spotted owl numbers, with losses at that point estimated at 3.8–8.4 percent annually. These findings suggest that barred owl range expansion may have been a significant and perhaps dominant, but then overlooked, factor in spotted owl decline in the 1980s and earlier.

Grand Staircase-Escalante

The Grand Staircase-Escalante region had attracted the attention of geologists since the late 1940s. A part of the 1.9 million acre area had been found to contain vast coal deposits, and by the early 1990s plans were being made to establish coal mining operations over a portion of the land surface. However, groups of environmentalists were following these developments and in 1995 they began soliciting help at the federal level to stop development. Thus, when in the fall of 1996 President Clinton designated Grand Staircase-Escalante as a national monument, the move was hailed by many as a victory for the environment.

But did the environment really benefit from the monument designation? Natural systems contained within monument boundaries undoubtedly did. But what about the environment as viewed from a broader perspective? Upon reflection, there is little cause for celebration. In addition to the vast coal deposits in Grand Staircase-Escalante, the area is estimated to contain vast amounts of other mineral resources. These include an estimated three million tons of zirconium/titanium, as well as deposits of petroleum, natural gas, sand and gravel, phosphorous, silver, copper, lead, zinc, and molybdenum. Many of these materials were on the U.S. net import list in 1996, and two that weren't on that list are today supplied by net imports. Thus, in essence what happened in the Grand Staircase-Escalante declaration is that the United States took yet one more step away from taking responsibility for its own consumption, and in the process further shifted the environmental impacts of mining to regions outside the U.S.

Net Import Reliance for Key Minerals
Contained Within Grand Staircase-Escalante

MINERAL	NET IMPORTS 1996 (% OF DOMESTIC CONSUMPTION)	NET IMPORTS 2015 (% OF DOMESTIC CONSUMPTION)
Petroleum	52	27
Zinc	35	82
Natural gas	16	4
Uranium	83	80+
Silver	net exporter	72
Copper	12	36
Phosphorous	net exporter	5–10
Lead	14	31
Vanadium	<10	100
Molybdenum	net exporter	net exporter
Titanium	net exporter	68
Titanium mineral concentrates	63	91
Zirconium	40	23

In announcing the national monument declaration, no mention was made of ongoing massive importation of the minerals involved or of any plan for increasing domestic recovery or recycling of these minerals, or of any examination of what the environmental impact might be if raw material importation were to increase as a result of this declaration. The similarity to the 1992 Clinton Forest Plan is striking.

What might President Clinton have done in this case? He might have taken advantage of his 1996 Grand Canyon photo opportunity to announce formation of a commission to examine increasing regulatory barriers to mining activity in the U.S. He might have initiated a national dialogue on consumption, or on finding ways to take greater responsibility for our consumption. Or he could have begun an effort to position the U.S. as a world leader in metals recovery and recycling. He might also have taken more seriously the request by mining companies to allow mining activity to proceed in one corner of the area.

Within several years of the Grand Staircase-Escalante, then Secretary of the Interior Bruce Babbitt issued moratoriums on minerals development in several known resource-rich areas and introduced regulatory changes, including an amendment to mining regulations. His actions, in turn, served to increase the risk and expense of mining ventures within the U.S.

Roadless Designation for Western Forest Lands

Eight days before leaving office in 2001, President Clinton designated 58.5 million acres of federally-owned forest land in the West, an area equivalent to over 2.7 percent of the land area of the U.S., as off limits to road building. In making the declaration, the outgoing president was acceding to the desires of a number of environmental organizations that had long pushed for prohibition of roads so as to effectively add vast areas to the federal wilderness system, and to make these areas economically unattractive as potential sites for mining, drilling, and timber harvesting activity.

The quest for protection of vast roadless areas had begun years earlier with establishment of the National Wilderness Preservation System by President Lyndon Johnson in 1964. By 1971, the system encompassed about 10 million acres in 89 distinct wilderness areas. Another 4.5 million acres were contained within designated primitive areas. However, the area of potential wilderness was much greater. In 1970, the USDA-Forest Service identified another 36 million acres of roadless, undeveloped federally managed land and yet another 54 million acres of such land that was part of the national park system and national wildlife refuges.

In early 1971, the Sierra Club issued a report that called for expansion of the federal wilderness system such that the land area encompassed by the "two extremes of land utilization—wilderness and total development—should be approximately equal." Looking forward, it was proposed that an area of land twice the area of total development in 1964 be set aside in wilderness reserves. The area of total development in 1964 was estimated at 55 million acres. In essence, the report called for designation of all existing primitive areas, and roadless and undeveloped areas, as wilderness. Pressure was then directed toward President Richard Nixon to act upon the recommendation.

Nixon acted within months to request additions to the national wilderness system, signing into law nine bills affecting 912,000 acres. He

was reluctant, however, to accede to requests for wilderness designation of vast land areas, in part because of opposition from the minerals,
forest products, and energy industries, but also because the question
of where the nation's future raw material supplies would come from
was at the time very much in the forefront of thinking among legislative and industry leaders. The National Materials Policy Act had been
passed into law by Congress less than a year before the wilderness expansion request, and a National Commission on Materials Policy formed.
Among the observations and recommendations that appeared in the
first report of the Commission were:

*The loss of access to timber and minerals [that would occur
from wilderness classification] on National Forest Lands may
be a serious deprivation . . .*

*To promote the search for scarce materials, the Federal
Government should facilitate access to public and private lands.*

Thus, at precisely the time when a twelve-fold increase in the size of
wilderness reserves was being sought, many were questioning whether
the nation was adequately positioned to provide future supplies of basic
raw materials and were calling for more, not less, access to public lands.
Nixon maintained his support for establishment of additional wilderness,
but only for modest additions. He subsequently approved additional wilderness designations involving 1.3 million acres.

Subsequent presidents—Gerald Ford, Jimmy Carter, Ronald Reagan,
George Herbert Walker Bush, and Bill Clinton each made additions to
the federal wilderness system. By far the largest addition was the wilderness designation by President Carter of over 100 million acres in Alaska.
By the end of 2000, the area of designated wilderness in the lower 48
totaled 47 million acres. The total would likely have been much greater
had President Clinton been able to gain the support of a Republican-
dominated Congress. Lacking such support, Clinton issued a presidential
proclamation in October 2000, permanently prohibiting road building on
58 million acres. The move effectively shifted these acres to a wilderness

status, and likely set the stage for an official wilderness designation at a later date.

It was seemingly a victory for the environment. However, as with other environmentally-inspired actions of the Clinton administration, the roadless designation was not accompanied by any action designed to curb or moderate consumption of the resources that these lands are known to harbor. Nor did the President use the occasion to remind people that minerals, fossil fuels, and timber must come from somewhere, and that the designation meant that other regions would have to be tapped for raw materials development in the future. Consumption-related concerns were also not voiced by any of the environmental organizations supporting the roadless designation.

Unintended Consequences

The environmental community is well aware that ecosystems are complex, and that the web of life interconnects, linkng both directly and indirectly the health and well-being of organisms at all levels of the food chain. There is apparently less appreciation for the fact that other systems, including the global market system, are also quite complex and interconnected.

A sharp decline in softwood harvesting in the Pacific Northwest was viewed in the early 1990s as a top environmental priority for the environmental community of that region based on concern for the northern spotted owl and the marbeled murrelet. But, as accurately predicted by University of California professor William McKillop, success in reducing harvests triggered a number of undesirable and entirely predictable developments, that included shifts in harvest activity to less productive land with far less environmental oversight than exists in the U.S.

The spotted owl recovery plan also triggered a number of consequences that were apparently unanticipated by those drafting alternatives for the President's consideration. An immediate consequence was that wood began to flow into the Pacific Northwest from Idaho, Montana, and as far eastward as the Black Hills of South Dakota, and from New

Zealand, Chile, and Uruguay. Other countries also took advantage of the changing market. Then, within two years of implementation of the plan, it was reported in Portland's daily newspaper *The Oregonian* that increased importation of softwood logs had increased risks to Oregon's forests due to the possibility of introducing populations of exotic insects. And, as earlier indicated, in the southern U.S., another location to feel increased pressure for harvesting because of developments in the Pacific Northwest, softwood harvest levels began to exceed net growth for the first time in more than 65 years.

Another (perhaps) unintended consequence of the spotted owl recovery plan was the loss of forest harvesting and wood products manufacturing infrastructure in several regions, a loss that today serves to preclude cost effective approaches to addressing forest health issues. In addition it led, as previously noted, to a sharp increase in the balance of trade deficit traceable to wood and wood products consumption.

To the credit of President Clinton, he did, in his first year in power, create a Council on Sustainable Development, which through the work of its Population and Consumption Task Force, called attention to consumption and the role of population growth as a driver of consumption growth. This work, however, was never followed up on by the President. He also proposed a number of initiatives focused on energy consumption. Proposals related to conservation and development that appeared to be based on recognition of the need for ongoing supplies and the importance of increasing domestic production of energy were drafted. Unfortunately, the administration was unable to gain acceptance and funding of most of what was advocated in a Republican-dominated Congress.

What Is an Environmental Leader Anyway?

In the years since Clinton left office his environmental legacy has faded somewhat. Nonetheless, he appears on every top ten list, ranked number four by *Good Housekeeping* and *Green Guide,* and number eight by the *New York Times* and also in a survey of twelve leading environmentalists and environmental groups.

In thinking about the characteristics that define an environmental leader, it is useful to consider how an anti-environmental leader might act. Interestingly, popular perception suggests that the very next U.S. leader to succeed Mr. Clinton—President George W. Bush—is a perfect example of one who exudes anti-environmental values. In fact, the 2000 election of G. W. Bush to the nation's highest office was viewed within the environmental activist community as nothing less than an utter disaster. As a conservative Republican with known ties to the oil and gas industry, and with a Vice-President who fit the same description, Bush was viewed as beholden to large corporate interests who wanted to, in the words of more than one writer, "gut the nation's environmental laws." Robert F. Kennedy, Jr. offered a blunt assessment of Mr. Bush in 2005, stating flatly that "This is the worst environmental president we've had in American history."

So what exactly is it that defined President Bush's so strongly as anti-environmental in the minds of many Americans? The actions listed most prominently by the President's critics include the following:

- During his first days in office he sought to undo some of the national monument designations that President Clinton made as his second term was winding down.
- He sought to overturn the roadless designation of over 58 million acres of federal forests in the western United States.
- He pushed for and achieved passage of a Healthy Forests Restoration Act that opened millions of acres of federal forests to logging for the purpose of thinning forest stands and made legal appeals of this kind of activity more difficult than previously.
- Through May 2006 (by which time he had already been singled out as the worst environmental President), he had not proposed a single new national park or created any new national monuments. (However, the following month, Mr. Bush established the Northwestern Hawaiian Islands Marine National Monument, the world's largest marine conservation area encompassing nearly 140,000 square miles).
- He signed into law only minimal additions to the federal wilderness system.

- He pushed for modification of the Endangered Species Act and issued a rule change to roll back its requirements six weeks before leaving office.
- He was known to favor extraction of oil and gas from the Arctic National Wildlife Refuge (ANWR).
- He advocated a renewed emphasis on production of nuclear energy.
- He was very late in advocating energy conservation as part of a national energy strategy and even then only half-heartedly proposed funding such efforts.

It is not difficult to understand why G.W. Bush's record is viewed as anti-environment. However, viewed from a different perspective this record is not as damning as often described. For instance, many of Mr. Bush's actions could be interpreted as no worse than those of his predecessor. Recall that Mr. Clinton took a number of steps to forestall raw materials development without in any way addressing consumption. Mr. Bush also didn't address consumption, but did on the other hand act consistently to seek greater domestic development of raw materials. While perhaps not his intent, his proposals pointed toward acceptance of greater responsibility for domestic consumption, and as a consequence, reduction of environmental impacts of consumption outside the borders of the U.S. We should carefully consider whether one is necessarily worse than the other from an environmental point of view. It is also worth considering what the reaction of various interest groups is likely to be to *any* leader who proposes greater domestic raw materials extraction.

Before leaving the topic of Mr. Bush's environmental record, it should be noted that just two weeks before leaving office, the President invoked the Antiquities Act to establish three additional marine monuments encompassing 125 million acres of habitat in America's Pacific Atoll. When added to the earlier designation in Hawaiian waters, Bush's actions amounted to the most sweeping use of the Antiquities Act since enactment of this law in 1906.

In rankings of environmental Presidents, such as those cited previously, six names other than that of Bill Clinton commonly appear—

Teddy Roosevelt, Franklin Roosevelt, Richard Nixon, Lyndon Johnson, Jimmy Carter, and Barack Obama. What did each of these Presidents do to achieve an environmental legacy?

Teddy Roosevelt created the first national wildlife refuge in 1903 through the issuance of an executive order, and subsequently created 52 more national refuges by the end of his term. He moved the U.S. Forest Service from the Department of the Interior to the Department of Agriculture with a goal of facilitating wise use. And, he signed into law the American Antiquities Act that established presidential authority to create national monuments.

Franklin Delano Roosevelt created the Civilian Conservation Corps which built hiking trails, planted billions of trees, cleaned up streams, and constructed more than 800 parks across the U.S. While aimed primarily at creating jobs in a time of economic crisis, the Civilian Conservation Corps left a lasting environmental legacy, including what would become a national network of state parks.

Richard Nixon's tenure was marked by passage of the Endangered Species Act (1969), the National Environmental Policy Act (1970), the Clean Water Act (1972) and the Safe Drinking Water Act (1974), establishment of the Environmental Protection Agency (1970), strengthening of the Clean Air Act (1970), and the addition of over 2.4 million acres to the federal wilderness system. Nixon was also the only President in history to attempt to initiate a national dialogue on population growth.

Lyndon Johnson's presidency marked passage of the Wilderness Act of 1964 (and in the process designated 9.1 million acres as wilderness), the Land and Water Conservation Act of 1965, the Endangered Species Preservation Act of 1966, and the National Trails System Act of 1968.

Jimmy Carter vigorously promoted energy conservation through appeals to citizens to turn thermostats down in winter and up in the summer. He is also the only modern-day President to address consumption in general, doing so by using a national television broadcast to challenge what he described as self-indulgence and over-consumption on the part of many Americans. The effort backfired and was seen by many citizens as blaming them for a number of ills during a tumultuous time in U.S.

history. However, the one act that indelibly marked his presidency in the eyes of environmentalists was his approval of the Alaska National Interest Lands Conservation Act (1980) that set aside 104 million acres, or more than one-fourth of the State of Alaska, as part of the federal wilderness system.

Barack Obama compiled an impressive environmental resume that included decisive actions to increase vehicle fuel standards; increase investment in clean energy; limit emissions of mercury, soot, and smog-forming chemicals; establish targets for reduction of carbon emissions by the federal government; establish targets for limiting carbon emissions more broadly; engage China in mutual setting of carbon emissions reductions targets; and to encourage energy conservation within and outside of the federal government.

President Obama set aside more land and water than any of his predecessors—over 264 million acres—primarily through National Monument designations. One of the most notable was action to increase the size of the Pacific Remote Islands Marine National Monument to almost 490,000 square miles, expanding a monument originally created by G.W. Bush by about 258 million acres. But Obama's administration also put into place new rules for management of federal forest lands, viewed widely as pragmatic and a significant improvement over rules that were put into place three decades earlier. His administration also stayed the course in the face of pressure to place a moratorium on leasing public land to mining firms. In addition, during his administration, federal agencies aggressively promoted and invested in bioenergy and biofuels, and took steps to achieve greater wood use in federal building projects.

By early 2015, several had suggested that Mr. Obama was *the* environmental President. Navin Nayak, senior vice president of the League of Conservation Voters was effusive in his praise, saying that "The president has been the greenest president we've ever had. In the context of comparing his accomplishments to past presidential records, it's not even close." In late 2014 the American Political Journalism Organization *Politico* remarked of Obama that "He didn't set out to be an environmental president. He is now."

Subsequently, Mr. Obama took action to block the Keystone pipe-
line, a move that won additional praise from environmental advocates,
but he subsequently also developed a plan to allow offshore drilling along
the Atlantic coast. Predictably, the latter move sparked vigorous protests
from both environmentalists and local governments all along the Atlantic
seaboard, with local opposition focused largely on potential adverse impacts on tourism and recreation. Ultimately, the President reversed his
earlier position and opted for a drilling ban.

National Monument designations in 2016, amounting to about 1.8 million acres of California desert and 1.6 million acres in Utah and Nevada,
were accompanied by a White House release which proclaimed that the
actions demonstrated "the Administration's strong commitment to aggressive action to protect the environment for future generations." There
was no mention of potential environmental impacts in other countries or
of current global resource concerns, although previous statements of Mr.
Obama—most notably his 2008 inaguration speech—suggest a clear understanding of the potential for international transfer of impacts linked to
domestic consumption.

Other than Franklin Roosevelt, Nixon, and Obama, the defining acts
that brought others on the top environmental Presidents list favorable attention in the eyes of environmentalists were the designations of large
tracts of lands as parks, preserves, and reserves. There is no indication
that any of those on the list gave any serious consideration to potential
environmental implications of their actions on regions outside of U.S.
borders. While Richard Nixon did not act to set-aside large areas of land
from development, he is recognized for the many forward-looking environmental initiatives that were signed into law during his tenure.

Although the examples provided in this discussion all center on only
several Presidents, actions to limit or halt resource extraction locally,
while completely ignoring raw material needs linked to consumption,
occur in the United States on a daily basis, at all levels of government,
and involving citizens of every political persuasion. Can the U.S. continue to act in this way from an economic point of view? Are such actions as outlined above ethically defensible? Socially defensible? Is

environmental policy that ignores consumption really beneficial to the global environment?

It is worth considering what the standards for an "environmental President" ought to be. In fact, serious re-examination of what it is that defines an environmental leader, whether working in government, environmental non-governmental organization, or business and industry, is long overdue.

References

Allison, M., R. Blackett, T. Chidsey Jr., D. Tabet, R. Gloyn, and C. Bishop. 1997. *A Preliminary Assessment of Energy and Mineral Resources within the Grand Staircase-Escalante National Monument.* Utah Department of Natural Resources, Utah Geological Survey Circular 93, January.

Alter, L. 2015. "Who was the Greenest President? 12 Environmental Groups are Polled and the Results Might Surprise You." *Treehugger,* Feb. 16, Accessed August 15, 2015.

Barry, J. 2001. "Clinton's Conservation Legacy: Can Bush Overturn It?" *Sierra Club Newsletter,* March.

Booth, W. 2004. "A Slow Start Built to an Environmental End-Run; President Went Around Congress to Build Green Legacy." *The Washington Post,* February 22.

Bryce, E. 2012. "America's Greenest Presidents." *New York Times,* Sept. 20.

Buchanan, J. 2016. Periodic Status Review of the Northern Spotted Owl in Washington. Washington Department of Fish and Wildlife.

Burns, J. 2015. Spotted Owls Still Losing Ground in Northwest Forests. Oregon Public Broadcasting/EarthFix. December 10.

California Natural Resources Agency. 2016. Status Review of the Northern Spotted Owl in California. Department of Fish and Wildlife, Report to the Fish and Game Commission, Jan. 27.

Chalt, J. 2013. "Obama Might Actually be the Environmental President." *New York Magazine,* May 5.

Chemnick, J. 2012. "CLIMATE: Does Obama's Record Measure Up to Leadership Credentials Touted in Party Platform?" *Environment and Energy Daily,* September 12.

Clinton, W. 1996. *Remarks by the President in Making Environmental Announcement Outside El Tovar Lodge, Grand Canyon National Park, Arizona.* The White House, Office of the Press Secretary, September 18.

Green Guide. 2014. "The 7 Greenest Presidents in the History of America," Jan. 18, Accessed August 14, 2015.

Hamilton, N. 2010. *Presidents—A Biographical Dictionary, 3rd ed.* Revised by Freidman, I, 374.

Hendershot, D. 2001. "A Presidential Role in Preserving Public Land." *Smoky Mountain News,* February 1.

Howard, B. 2015. "The 10 Greenest Presidents in U.S. History." *Good Housekeeping,* Jan. 4.

Howard, J. and Jones, C. 2016. U.S. Timber Production, Trade, Consumption, and Price Statistics, 1965-2013. Research Paper FPL-RP-679. USDA-Forest Service, p. 15.

Hughes, J. 1999. "Carter, Nixon Rank High as Environmental Presidents." *Seattle Daily Journal* of Commerce, December 28.

Kohler, V. 1992. "Plant Experts Warn Against 'Hitchhikers'." *Oregonian,* August 12.

Linden/Yakutsk, E. 1995. "The Tortured Land." *Time* 146, September 4.

McKillop, W. 1995. "Industrial Forestry and Environmental Quality." S.J. Hall Lecture, University of California, Berkeley.

Menzie, D. 2005. Statement before the House Resources Committee, Sub-committee on Energy and Mineral Resources, March 16.

Menzie, D. 2006. Testimony before the Committee on Resources Subcommittee on Energy and Mineral Resources, United States House of Representatives, Hearing on Energy and Mineral Resources for Development of Renewable and Alternative Fuels Used for Transportation and Other Purposes, May 18.

Murray, P. 1996. Record of a statement to the U.S. Senate, March 13. Congressional Record Volume 142, No. 35.

National Commission on Materials Policy. 1973. *Materials Needs and the Environment Today and Tomorrow.* Washington, D.C. (June)

Samuelsohn, D. 2014. "The Greening of Barrack Obama." *Politico,* November 18, Accessed August 14, 2015.

Wapner, P. 2001. "Clinton's Environmental Legacy." *Tikkun,* March/April, Accessed August 14, 2015.

PART 5

CHANGING COURSE

This final section examines some of the things that might be done to embark on a course for meaningful change. Examined briefly as well is what continuation on the current path might mean for future generations.

Finally, key points from preceding chapters are summarized. Questions are posed for those who wish to consider further the issues discussed.

CHAPTER 12

Rethinking Consumption

"Infinite growth of material consumption in a finite world
is an impossibility."

—E.F. SCHUMACHER

Within the span of one human lifetime, the world population has tripled. In the same time frame, the global economy has expanded by a factor of 25, consumption of basic resources by a factor of five, and energy consumption by a factor of more than seven. Nations that were dominated only several decades ago by centuries-old practices and backwater economies have in short order joined the 21st century. The relative wealth of nations and the relationships between them has changed as well. Economies that were once highly dependent on the economic health and consumptive patterns of a relatively few highly developed nations are today far less dependent. Concurrently, former raw material suppliers are becoming competitors for those same resources.

All of this has brought unprecedented demands on the Earth's bounty and the natural systems that support life. Some worry about raw material exhaustion, others about distribution of resources and potential for conflicts. Many are concerned about impacts on the environment and whether the present course can be sustained. In this regard, the situation is one in which the capacity of Earth and its natural systems to provide and adapt is unknown, but where shrinking margins for error are a reality.

Looking forward, during the period 2015–2050 more people are expected to be added to the population than the total of humanity in 1940. Continuation of current trends is expected to result in annual consumption of basic resources that are three times that of 2000 by as soon as

2050. Specifically, business as usual is predicted to result in more than a doubling of biomass consumption, a near quadrupling of fossil fuel consumption, and a tripling of the annual consumption of metals and construction minerals.

Increasing numbers of analysts believe that the current rate of consumption globally is unsustainable in the long term. If current levels of consumption are problematic, then those of the near future could spell trouble, since population and economic growth and consumption trends suggest very large increases in consumption and resource extraction in the decades ahead.

Given these concerns, consumption patterns in the most economically developed countries are now in the spotlight. Annual per capita resource use in these countries—the U.S., Canada, Western European countries, Japan, and Australia—is double the world average of 8–10 metric tons per capita, and four to five times that of the poorest countries. U.S. per capita consumption of raw materials is, incredibly, almost double that of the other most developed countries.

Consumption, Environment and Equity

Global concern about impacts of rising consumption first surfaced in the early 1970s, several decades into the explosive economic and consumption growth that followed World War II. Continued growth of both world population and economy, and accompanying consumption, soon led to greater and more widespread concerns about not only the environment and long term sustainability, but also about global equity. There is major apprehension about these issues today.

Discussion of what might be done to address such concerns inevitably returns to concepts outlined in the Brundtland Commission report, and the publication *Caring for the Earth: A Strategy for Sustainable Living,* a joint report of the World Conservation Union, the United Nations Environment Programme, and the World Wide Fund for Nature. Both reports were published in the early 1990s, and both chronicled rising environmental impacts linked to increasing resource extraction and consumption, while identifying needs for change in the global dynamic. Both also recognized the link between poverty and high birth rates, and the role that improved economic conditions can play in reducing the rates of population growth.

The *Caring for the Earth* report focused on the daunting challenge of bringing about more economic growth in large parts of the world while simultaneously reducing or moderating global environmental impacts. Included in this report were the following observations:

- A relative few, most of whom live in high-income countries, enjoy high living standards, and consume a disproportionate share of the world's resources.
- The well-off minority may support reduced resource consumption through increases in efficiency, and perhaps even stabilization of living standards, but are not likely to willingly support reduction of standards of living.
- The majority of people throughout the world, mostly but not all living in lower-income countries, have far lower standards of living, use a far lower proportion of the world's resources, and in many cases suffer from the diseases of poverty.
- Modern communications and tourism have brought the luxury of the rich before the eyes of the poor, and the latter no longer accept disparities with patience or as part of some natural historical order.

A key takeaway here is it has long been recognized that residents of the most developed nations are unlikely to voluntarily reduce their consumption *per se,* but might support a strategy to reduce resource consumption through increases in resource use efficiency. Another is that residents of developing countries aspire to greater levels of affluence and consumption—something that is rapidly becoming reality.

In 1993, fourteen years after President Carter openly challenged self-indulgence and over-consumption of many Americans, President Clinton's Council on Sustainable Development also called attention to consumption and the role of population growth as a driver of consumption growth. The report highlighted high levels of consumption relative to much of the world and provided recommendations for reducing consumption. Subsequently, the UN Development Program's 1998 Human Development Report was dedicated to the question of consumption and sustainability, pointing out large disparities in global consumption patterns and raising questions about global equity.

In 2000, the well-publicized finding that an average child born in the United States would consume 30 to 50 times more resources in his or her lifetime and have a similarly greater impact on the global environment than a child born in a developing country again brought attention to consumption and global equity. In that same year, the authors of the United Nations Environment Programme publication *Global Environmental Outlook—2000* took a strong position on consumption in developed nations. They called for *a tenfold reduction in resource consumption in the industrialized countries,* arguing that this was necessary if adequate resources were to be released to supply the needs of the developing countries.

More recently, the World Resources Forum has estimated that by 2050 natural resource use globally needs to be brought into the range of 5–6 metric tons per person per year, while emissions of greenhouse gases should be limited to no more than two tons per capita. In comparison, North American per capita annual resource use in 2008 was about 21 metric tons. This figure rises to 38 metric tons when materials used in producing imported goods are included in the calculation. U.S. greenhouse gas emissions in the period 2011–2015 were about 17 metric tons per capita, about three times the global average of 4.9.

The view that efficiency improvement in materials use is the key to solving this problem remains dominant. This translates to implicit adoption of technology as a primary strategy, with a specific focus on increasing raw material use efficiency, and vastly improving recovery and recycling. Some, however, are convinced that technology alone won't be enough, and that high per capita consumption within developed nations is a problem that will have to be addressed directly at some point.

Tiptoeing Around Consumption

Ironically, the revelation that high consumption is associated with large and adverse environmental impacts has received little attention in the U.S.

and other developed nations. Indeed, in many circles developed countries are seen as the vanguard of environmental protection because they were (and are) also generally the most environmentally pristine. To some, efforts to draw attention to consumption issues are seen as simply part of a bigger plan to re-engineer societies of the developed world.

Governments periodically encourage consumption as a means of stimulating the economy. However, the notion of reducing consumption is another elephant-in-the-room topic, generally ignored in seeking possible solutions to environmental or raw material concerns. Few political leaders in the developed countries want to suggest that people should be happy with fewer consumables. In fact, no nation state has ever voluntarily reduced its standard of living (i.e., its consumption), suggesting that reducing societal consumption per se might be extremely difficult to achieve, and that even discussing such an idea might amount to political suicide.

As discussed previously, it is for this reason that by the mid-1990s the environmental community recognized that although consumption leads to a host of negative environmental impacts, a strategy of seeking to reduce individual consumption within the most economically developed nations was not likely to succeed. Consequently, no serious effort to do so has ever been mounted.

In a chapter entitled "Confronting Consumption," in a book of the same name, the reticence of North Americans to seriously address consumption was examined. The authors observed that consumption and consumerism tend to be relegated to polite talk, if discussed at all, since discussion of these topics can quickly become awkward as discussion of any depth can lead to questions about deeply held convictions about such things as lifestyle, measures of success, personal choice, and related matters. They described as "not surprising" that society has settled on rather comforting concepts such as "sustainable development" and "green consumption" in lieu of more substantive attention to ever-increasing consumption, its roots, global impacts, and implications for the future.

Pope Francis, in his 2015 encyclical on the environment, is sharply critical of actions of the high consuming countries and failure to address consumption levels, saying:

*To blame population growth instead of extreme and selective
consumerism on the part of some, is one way of refusing to
face the issues. It is an attempt to legitimize the present model
of distribution, where a minority believes that it has the right
to consume in a way which can never be universalized, since
the planet could not even contain the waste products of such
consumption.*

In making this statement, Francis dismisses the notion that population
growth is an environmental issue except for the effects of geographic imbalances in population density. It is an interesting perspective. Nonetheless,
his criticism of high consumption and resulting social inequity is quite
harsh.

An overriding view across society at large, then and now, is that vigorous protection of the environment and economic growth are competing
objectives. Langdon Gilkey, a theologian and prolific author, described
the situation this way:

*Something must be done. Again, however, as with social justice
(and gender), the main problem is not in devising solutions;
there is much that can be done and done immediately. The
problem is once more that any resolution hurts. Every solution
means a sacrifice of some sort: perhaps it will raise production
costs, and so jobs will be lost. As Robert Heilbroner gloomily
prophesied, no society will voluntarily lower its gross national
product or its standard of living. There are no solutions that are
cheap, none that are free. Hence policies that save trees, clean
the air or the water, reduce toxic waste, rescue the ozone layer—
whatever—these policies all cost something precious, they hurt
someone and perhaps they hurt all of us. Good policies require
sacrifice. Thus, whether it be a capitalist or a communist society,
these policies tend to be ignored, their evidence disputed, other
experts called in. And so we delay, and in the end we endanger
our world, our children's future, our species itself.*

Developments vis-à-vis consumption have not played out as anyone forecast just four decades ago. On the one hand, global consumption has literally exploded. On the other hand, although no deliberate efforts were undertaken to address global inequity, a reordering occurred nonetheless. As more and more countries experienced rapid economic growth through the late 20th and early 21st centuries, issues of global inequity faded a bit, aided in part by the deep U.S. economic recession of the early 21st century and the resulting decline in U.S. resource consumption. The U.S. 2007–2010 recession, combined with concurrent economic woes within Europe and Japan brought about, if only temporarily, sharp reductions in developed-country consumption. Yet, as the U.S. economy began to rebound from recession, and as economic growth in China and other rapidly developing nations continued at a brisk pace, the rate of increase in global resource consumption began rising again.

Technology and Consumption

Technological advances are often referenced as the answer to almost any problem. But that has not always been the case. In 1974, biologists Paul Ehrlich and John Holdren published the I = PAT equation, to explain the impact of human activity on the environment. According to the equation, impact (I) equals population (P), times affluence (A—a proxy for consumption), times Technology (T—defined in this case as the influence of technology on procurement of resources and transformation into useful goods and wastes). The equation was widely criticized for suggesting that improved technology leads to increased environmental impact, and for being overly simplistic. Its developers and others have responded that although better technology can lead to reduction of environmental impacts, the net effect is often increased overall impact on the environment.

In 1996, IPAT was succinctly summarized in an article entitled "Can Technology Save the Earth?" Two basic arguments were identified as weighing against the influence of technology, one being that "technology's success is self-defeating" in that "if we solve problems, our population grows and creates further, eventually insurmountable problems."

The second argument identified is that technology also produces guns and bombs and is a force for evil as well as good. The criticism didn't stick. In 1997 the authors of the international best seller *Factor Four: Doubling Wealth–Halving Resource Use* explained "that it is possible to do far more with less, a goal that is not the same as doing less, doing worse, or doing without." It was this publication, more than any other, which shifted the promise of technology back to center stage as a strategy for dealing with the population/consumption/environmental impact dilemma.

Technology was soon linked to a new concept—that of decoupling. The decoupling concept was succinctly explained in 1999 by Yukiko Fukasaku of the Organization for Economic Cooperation and Development:

> *It used to be taken for granted that economic growth entailed parallel growth in resource consumption, and to a certain extent, environmental degradation. However, the experience of the last decades indicates that economic growth and resource consumption and environmental degradation can be decoupled to a considerable extent. The path towards sustainable development entails accelerating this decoupling process.*

Decoupling (also sometimes referred to as dematerialization) is all about providing goods and services with more efficient use of physical material, creating in the process reduced materials consumption and lesser quantities of emissions. Fundamentally, the idea is to do more with less.

Examples of success in dematerializing are many. Competitive pressures provide an incentive for business and industry to improve materials use efficiency and reduce waste. Cost containment has led to a plethora of products manufactured at reduced cost using less material. Products across almost every industry are today less material intensive than just a few decades ago, the result of ongoing research and technology development driven by economics and free market competition.

Although proponents of decoupling are careful to point out that this does not mean reducing consumption, a central assumption in as-sessments of potential success of the decoupling movement is that the decades-long shift of populations from rural areas to cities will continue and that this will result in "more compact living." Seemingly imbedded in that assumption is that those living in cities will occupy less land, will therefore perhaps have smaller or non-existent lawns and less need for associated equipment, have greater access to public transportation and less need for private vehicles, and so on. Thus, it appears that at least some of the promise of dematerialization is that developed country consumption might decline through changing preferences and trends in succeeding generations.

One reason for focusing on technology and not on reducing con-sumption as the dominant strategy for protecting the environment is the complexity of the world in which we live. Progress across a wide array of environmental concerns requires simultaneous successes in a number of arenas. The challenge is well defined by the demographic transition model. Long-observed trends in developing countries suggest that envi-ronmental progress may actually begin with industrialization. The tran-sition from high birth and death rates to low birth and death rates, which result in lower and more stable rates of population growth and increased attention to education and environmental issues, is closely and positively linked to economic development. Somewhat ironically, progress toward greater sustainability in a number of countries has been observed to in-volve a shift to greater industrial activity, higher consumption, and some-times even a brief increase in the population growth rate. However, the higher incomes that industrialization tends to foster allow improvements in health care, better education, and job opportunities for women, factors that often lead to greater self-determination for women, reduced birth rates over the long term, and attention to environmental matters that tend not to be addressed until basic human needs are met.

Because of these interdependencies, while a singular focus on reduc-ing consumption might succeed in reducing one factor in the IPAT equa-tion, other factors are likely to remain unaddressed.

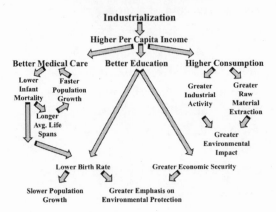

The Demographic Transition

The rationale for concentrating on decoupling economic growth from resource consumption was summarized in the publication *Cents and Sustainability: Securing a Common Future by Decoupling Economic Growth from Environmental Pressures*. A product of the Natural Edge Project, the volume builds on themes outlined in *Our Common Future* and presents a counter view to those who continue to argue for significant slowing of consumption as a means of addressing sustainability concerns.

Dr. Kenneth Ruffing, former Deputy Director and Chief Economist of the Organization for Economic Cooperation and Development Environmental Directorate, and Deputy Director of the UN Division for Sustainable Development, explained the need for a focus on decoupling this way:

> *Ultimately, environmental pressures can be reduced only by reducing the level of output (negative economic growth) or by transforming the economic processes that underpin growth in such a way as to reduce the pressures (emissions of pollutants, greenhouse gases and destructive use of natural resources) per unit of output. Since buoyant economic growth is a necessary, but by no means sufficient condition, for achieving most of the goals, the former is not an option. That leaves us with decoupling.*

Because environmental pressures are multiple, there is no single overarching solution—no magic bullet.

Challenges in Reducing Materials Consumption

Accelerating Technology Development

In seeking answers regarding materials use, technology seems to always bubble to the top. Some view current trends as simply an extension of the past. They see the situation resolving through ongoing technological development and associated investment. Others see very rapid, unprecedented change as a looming crisis, requiring an extraordinary rate of technological improvement such that concerted, coordinated effort and targeted investment in dematerialization will be needed to avoid environmental and social catastrophe. Still others doubt that technology will be sufficient to forestall lasting environmental damage and social upheaval.

Other than the UNEP/OECD dematerialization initiative, what steps are being taken to prepare for what might lie ahead? What strategies, for instance, are being pursued or considered within the most highly-developed nations?

Developed countries engage in and finance research dedicated to improving energy efficiency, reducing energy use in building construction and operation, developing alternative sources of energy, developing break-through technologies to increase materials and energy efficiency in basic industries, creating new lighter and stronger materials including advanced composites, developing renewable materials including bio-chemicals, and so on. This kind of activity has been ongoing within virtually all of the developed countries for decades.

Moreover, the private sectors of market-driven economies continually seek to reduce the quantity of resources used in each unit of goods produced, driven by competition and profit motivation. In industry after industry, the efficiency of raw material use has increased steadily, such that consumption of raw materials per unit of GDP has fallen markedly over many decades. This is a continuation of a trend that began long before regulatory pressures and governmental mandates. The combined effect of government involvement in materials use efficiency research and on-going process and product improvements in industry are a large part of the reason why materials use intensities (consumption of materials per unit of economic activity) have been falling for some time.

The good news here is that the institutions and funding mechanisms needed to mount a substantial research and development effort are already in place, as are precedents for working with private sector companies. The bad news is that all of this may not be enough. A key element in pursuing change is a renewed commitment to scientific investigation and technology development. The technological advances of the present are all based on fundamental and applied research of the past. If society has chosen to place all its bets on technology, then substantial and ongoing investment in R&D will be essential—perhaps substantially more than is presently the case.

In 2012 the European Commission indicated that the level of R&D funding needed for rapid change in technology development in Europe to be 3 percent of GDP. They found, however, funding levels to be far below that benchmark. In the period 2011–2014, only nine European nations invested more than 2 percent of GDP in R&D, and only five—Germany, Sweden, Finland, Switzerland, and Austria—invested more than 2.5 percent. For the EU-27 as a whole, R&D funding levels were 0.76 percent of GDP in 2012, about one-fourth that identified as needed to achieve rapid change. Globally, R&D investment as a percentage of World Gross Product was 1.8 percent in 2013. Major investors included the United States (2.8 percent), China (2.0 percent), Japan (3.4 percent), South Korea (3.6 percent), and Canada (1.9 percent).

Shifting Public Policy

Some of those who view technology solutions as insufficient to meet the challenge argue that government regulation through the use of various public policy measures might also be needed to achieve resource use reduction goals. Public policy instruments have long been used by governments to change consumer behavior. Among the available tools are regulation and mandates, incentives and disincentives (such as through subsidies and tax law), and facilitation of collaboration among levels of government and between government and the private sector. Governments can also play an important role in funding research, and in devising and supporting initiatives involving the private sector, universities, and others in pursuit of new and improved technologies and systems for materials use efficiency.

It is well established that tax policy can change consumer behavior. Ethanol incentives and cigarette taxes in the U.S. serve to promote alternative energy on the one hand, and discourage smoking on the other. Energy taxes are used in European countries to incentivize smaller vehicles and high building energy efficiency. The EU energy tax is a targeted form of a consumption tax that has been kept in place for decades through liberal and conservative governments alike, and is an important reason why per capita energy consumption in Europe is one-half that of the U.S.

In fact, differences between U.S. and EU energy taxation undoubtedly influence more than just energy consumption. Consider, for instance:

1) Why are U.S. homes in comparison to those of the E.U.:
   so much larger?
   so seldom designed so that zone heating could be effectively employed?
   so much more dispersed?
   so much less likely to be served by rapid transit?
2) Why are U.S. automobiles so large, and so fuel inefficient, in comparison on average to those in the E.U.?
3) Why do U.S. residents travel, on average, 2.5 times the number of auto miles annually per capita and three times the number of air miles, but only one-half the distance per capita by rail and bus transit systems?

In a word, a key part of the answer to all of these questions is energy, and more specifically cheap energy. A long history of abundant, low cost energy has allowed U.S. residents to make everyday decisions, large and small, with little or no thought to either the price or availability of energy. Faced with a decision whether to live close to work or in the distant suburbs with a long commute, the first and often only thought has been about the driving time. Energy efficiency or a hot tub? Who cares that the hot tub will likely require more power than any appliance? Compact or full size car? More often than not this kind of decision pretty much hinges on the initial price of a vehicle.

A consumption tax (sometimes referred to as a value-added tax) can be applied to consumption in general. These kinds of taxes are applied in all European countries, and also in Australia, New Zealand, Singapore, and Canada. The Canadian tax is called a goods and services tax (GST). Interest in a generally-applied consumption tax within the U.S. Congress goes back over 40 years, as various elected officials have searched for alternatives to the income tax. The first introduction of legislation in the U.S. to establish a consumption tax occurred in the 98th Congress in the mid-1980s, and similar legislation has been introduced in both the House and Senate in every subsequent Congress through the 114th.

Recent proposals have called for establishment of a federal consumption tax ranging from 10–23 percent on the purchase of goods and services, while at the same time eliminating income tax liability for most or all Americans and dramatically reducing or eliminating corporate income taxes. Envisioned is a fundamental change in the way the U.S. federal government raises revenue. Rather than taxing income, the new tax system would raise revenue by taxing consumption. Most proposals have included provisions for protecting low- and middle-income families from excess taxation through a rebate system.

Billionaire and entrepreneur Bill Gates, also a champion of the consumption tax idea, has stated that such a tax would be simple, efficient, and fairer to people at every income level. Author and economist Michael Graetz makes a strong case for taxation of consumption and eliminating income tax in his book *100 Million Unnecessary Returns*. Although none of its promoters point to the effect that such a tax might have upon individual consumption, the prospects are interesting. In effect, implementation of a consumption tax could make it possible for each taxpayer to determine the level of tax paid simply by modifying his or her behavior: If you think you are paying too much tax you can simply reduce your consumption!

Whether the consumption tax will ever see the light of day in the U.S. remains to be seen. However, given oft-expressed unhappiness with the current tax code and the prospect of eliminating it, it would seem that many might be open to this alternative.

In recent years, taxation of fossil carbon emissions have been proposed, in this case for the specific purpose of reducing additions to atmospheric carbon. The effect would be similar to a consumption tax in that the effect would be to increase the costs to consumers of fossil fuels. The idea is to provide an incentive to reduce fossil fuel consumption, while at the same time incentivizing acceleration of development and implementation of renewable forms of energy. This is again a concept that deserves attention and serious discussion.

Encouraging Voluntary Consumption Restraint

Voluntary Simplicity

Some see voluntary reduction of purchases and accumulation of material goods as a viable avenue for achieving reductions in societal consumption. Proponents challenge the notion that initiatives aimed at getting residents of high consuming nations to reduce their individual consumption are unlikely to succeed, or at least to succeed on such a scale as to markedly impact overall consumption levels. Prospects for reduction of personal consumption through what is known as the voluntary simplicity movement were outlined in the book *Confronting Consumption,* and expanded upon in the widely acclaimed *Prosperity Without Growth.* In the latter, it is argued that prosperity and consumption are not the same thing, and that it is possible to build thriving societies in the absence of growth. It is an interesting premise, as indeed economies designed to function well only in an environment of growth may encounter significant difficulty as world populations stabilize, and as growth in the numbers of consumers and household formations ceases.

In *Confronting Consumption,* author Michael Maniates discusses the premise and successes of the voluntary simplicity movement in considerable detail, but faults leaders of the movement for not focusing on big picture issues such as rules, norms, and practices that govern work and compensation in the United States—changes that he posits are essential if the voluntary simplicity movement is to lead to substantial and lasting change. It remains unclear as to the extent to which simpler, less consuming life styles might be adopted, and the overall impact on consumption absent majority buy-in or fundamental change in the underlying factors which drive consumption.

Improving Reliability of Product Information

While there are questions about the probability of voluntary simplicity leading to significant change, it is nonetheless obvious that actions of individual consumers are important in changing consumption patterns. Aside from the quantity of goods consumed and the source of raw materials used to support that consumption, *how* society consumes and makes consumption decisions are important determinants of the impacts of consumption. One aspect of this is paying attention to environmental labeling initiatives designed to bring reliable information to the consumer.

In the not-too-distant past, environmental product labels were almost universally suspect. Advertising claims and product labels commonly included such meaningless terms as "eco-friendly," "green," and "all natural." However, regulation and oversight, coupled with new initiatives to quantify and verify product attributes, have brought some integrity to product claims. In recent decades, two distinct developments have occurred:

- product certification backed by independent third-party verification has been introduced for many product categories, and
- life cycle assessment, a method of systematically evaluating environmental performance, has been made available to consumers in the form of low-or-no-cost user-friendly websites and software.

Certification

Programs such as the UL Label and the Good Housekeeping Seal of Approval have long been familiar to North American consumers. Since the early 1990s a number of consumer-oriented product certification programs have been developed, focused on specific product lines. Third-party certification programs today cover a wide variety of products and product attributes. Now available are products that are demonstrably formaldehyde free (building products, floor coverings, fabrics), that have low or no volatile organic compound emissions (adhesives, paints, finishes), are organic (food), produced without the use of child labor (clothing and other products), are harvested in an environmentally responsible manner (fish, timber) or produced and traded in a socially responsible manner (wide variety of consumer goods), and so on.

Certification programs can be quite extensive. For instance, certification of wood is based on third-party assessment of forest management and timber harvesting based on publicly available standards designed to ensure appropriate protection of flora, fauna, water quality, soil productivity, historic areas, old trees, and more; attention to the rights of indigenous people; attention to rights of workers; and the well-being of local communities. Through these kinds of programs, a wealth of reliable information is provided to consumers that was previously not available.

Life Cycle Assessment

More significant, and particularly with regard to a goal of consuming less, is the relatively recent development of life cycle assessment (LCA) information and tools. LCA yields information about a wide range of environmental impact measures for products or product components. Measured are such things as embodied energy (total energy required in producing a product), fossil fuel consumption, total resource use, water use, global warming potential, ozone depletion potential, emissions to water, and more. Assessment encompasses product production, use, and disposal.

Data collected are more or less precisely measurable. They are gathered and analyzed following a set of internationally recognized rules, resulting in analyses that can be reproduced and verified. The best available science is applied in determining the environmental significance of resource flows and emissions. The measurable, reproducible aspects of life cycle assessment are key to its usefulness in environmental reporting, as it ensures that assessment is systematic, science-based, and free of intuition, bias and unsubstantiated claims.

Illustrating that intuition is not a reliable basis for assessing comparative environmental impacts of various materials or products, are responses to questions asked of large audiences of architects and engineers in recent years. Asked, for instance, which type of construction results in greater environmental impact—concrete or steel—assuming creation of otherwise identical buildings, the vast majority of votes favor concrete as the low impact alternative. In fact, systematic assessment almost invariably shows steel structures to have lower impacts across a wide range of impact measures. Among these are lower embodied energy, lower fossil energy consumption, lower global warming potential, and lower overall materials use. When steel construction is similarly compared to wood construction, wood is clearly seen to have lower impacts across the same range of measures.

As of this writing, life cycle assessment is being used primarily by manufacturers in identifying ways to reduce environmental impacts linked to their products and processes, by purchasing departments of large companies, and by architects and engineers who design structures and specify building materials. In the future, life cycle assessment-based information in the form of product labels for a wide range of products will help to inform purchasing decisions of individual consumers.

Life cycle assessment and LCA-based environmental labeling, used in combination with product certification, provides extensive information about the impacts linked to a specific shipment of a specific product. LCA offers information essential to reliably designing or selecting lower impact, lower resource consuming products. However, this tool will be effective only to the extent that people use it. Similarly, product certification programs can be of great value, but effectiveness requires that consumers pay attention and make purchase decisions accordingly.

The Un-Shopping Card

One of the best guides for making consumption decisions is that produced by the Oregon State University Extension Service. Dubbed "The Unshopping Card" the wallet-sized card begins with the question "Do I really need this?" Then, assuming that the answer to the first question is "Yes" the would-be consumer is led through the following series of questions:

- Could I borrow, rent, or buy it used?
- Is it made of renewable or recycled materials?
- Is it recyclable or biodegradable?
- Is it worth the time I worked to pay for it?
- Is it over-packaged?
- How long will it last?
- If it breaks, can it be fixed?
- How will I dispose of it?
- What is its environmental cost?

These questions are not about doing without. They are about consuming smarter, considering environmental attributes, and satisfying needs while using less.

Looking Forward

Contemplating change is hard for almost everyone. The prospect of substantial change induces near-universal automatic resistance. But the world today is a much different place than the world in which the vast majority of people grew up—a statement that rings true even for those still in their early 20s. Confronting the future, and a world of ever-more people and rising global affluence, is likely to require adaptation to new ways of thinking, acting, and living. Limitless growth may require rethinking. Environmental impacts of resource use will doubtless become more important. And global equity is likely to become a larger and larger issue. How we consume and how much we collectively consume underlies all of these factors.

References

Alcott, B. 2010. "Impact Caps: Why Population, Affluence and Technology Strategies Should be Abandoned." *Journal of Cleaner Production,* 18(6):552–560.

Allenby, B. 2006. "Dueling Elites and Their Catastrophic Visions." *GreenBiz,* October 31. Accessed March 5, 2015.

Ausubel, J. 1996. "Can Technology Spare the Earth? Evolving efficiencies in our use of resources suggest that technology can restore the environment even as population grows." *American Scientist* 84(2):166–178.

Barbour, S. 2000. *The Environment.* Greenhaven Press.

Chertow, M. 2001. "The IPAT Equation and Its Variants: Changing Views of Technology and Environmental Impact." *Journal of Industrial Ecology* 4:13–29.

Clarke, R., ed. 2000. *Global Environmental Outlook—2000.* United Nations Environment Program. London: Earthscan Publications Ltd.

Eckermann, F., M. Herczeg, M. Mazzanti, A. Montini, and R. Zoboli. 2012. *Resource Taxation and Resource Efficiency along the Value Chain of Mineral Resources.* European Topic Centre on Sustainable Consumption and Production, Working Paper 3/2012.

Eggert, R. 1990. "The Passenger Car Industry: Faithful to Steel." In *World Metal Demand: Trends and Prospects,* edited by J. Tilton. Washington D.C.: Resources for the Future.

Ehrlich, P. and J. Holdren. 1974. "Impact of population growth." *Science* 171: 1212–1217.

Ekins, P., B. Meyer, and F. Schmidt-Bleek. 2009. *Reducing Resource Consumption: A Proposal for Global Resource and Environmental Policy.* Gesellschaft für Wirtschaftliche Strukturforschung mbH (GWS Environmental Consulting).

Fischer-Kowalski, M., M. Swilling, E. von Weizsäcker, Y. Ren, Y. Moriguchi, W. Crane, F. Krausmann, N. Eisenmenger, S. Giljum, P. Hennicke, P. Romero Lankao, A. Siriban Manalang, and S. Sewerin. 2011. *Decoupling Natural Resource Use and Environmental Impacts from Economic Growth- A Report of the Working Group on Decoupling to the International Resource Panel.* United Nations Environment Program (UNEP).

Fukasaku, Y. 1999. "Stimulating Environmental Innovation." In *Special Issue on Sustainable Development.* OECD STI Review 25(2): 47–64.

Gilkey, L. 1995. "Order and the Transcendence of Order." In *An Ethos of Compassion and the Integrity of Creation,* edited by B. Walsh, H. Hart, and R. VenderVennan. Christian Studies Today.

Gore, A. 1992. *Earth in the Balance: Ecology and the Human Spirit.* New York: Plume Publishing.

Graetz, M. 2008. *100 Million Unnecessary Returns.* Grand Rapids, MI: Integrated Publishing Solutions.

Grueber, M. and T. Studt. 2013. *2014 Global R&D Funding Forecast.* Battelle.

IUCN/UNEP/WWF. 1991. *Caring for the Earth. A Strategy for Sustainable Living.* Gland, Switzerland, 43–44.

Jackson, T. 2009. *Prosperity without Growth: Economics for a Finite Planet.* London: Earthscan.

MacCleery, D. 2000. "Aldo Leopold's Land Ethic—Is It Only Half a Loaf Unless a Consumption Ethic Accompanies It?" *Forest History Today* (Spring), pp. 39–42.

Maniates, M. 2002. "In Search of Consumptive Resistance." In *Confronting Consumption,* edited by T. Princen, M. Maniates, and K. Conca. Cambridge, Massachusetts: MIT Press.

Männik, K., K. Eljas-Taal, J. Angelis, and A. Rozeik. 2012. *Funding Research and Innovation in the EU and Beyond: Trends during 2010–2012.* Analytical Report 2012 produced under the Specific Contract for the Integration of the INNO Policy Trend Chart with ERAWATCH (2011–2012) Technopolis Group. November.

Nath, V. 2001. *White Collar Invasion: Developed Country Policies Leading to Environmental Degradation in South.* Virginia Tech University, Center for Digital Discourse and Culture.

Organization for Economic Cooperation and Development (OECD). 2004. *OECD Environmental Strategy: 2004 Review of Progress.*

Organization for Economic Cooperation and Development (OECD). 2010. *Material Resources, Productivity and the Environment: Key Findings.*

Pope Francis. 2015. "Encyclical Letter Laudato Si' of Francis the Holy Father on Care for our Common Home." The Vatican, May 24.

President's Council on Sustainable Development (PCSD). 1993. *Population and Consumption Task Force Report.* Washington DC: U.S. Government Printing Office (June).

Princen, T., M. Maniates, and K. Conca. 2002. *Confronting Consumption.* Cambridge, Massachusetts: MIT Press.

Purves, R. 2010. "Introduction." In: Smith, M., K. Hargroves, and C. Desha. *Cents and Sustainability: Securing a Common Future by Decoupling Economic Growth from Environmental Pressures.* Natural Edge Project, p. xxix.

Smil, V. 2014. *Making the Modern World: Materials and Dematerialization.* West Sussex, UK: John Wiley and Sons, Inc.

Strigel, M., and C. Meine. 2001. *Report of the Intelligent Consumption Project.* Madison: Wisconsin Academy of Sciences, Arts and Letters, 39 pp.

United Nations Development Program (UNDP). 1998. *Human Development Report 1998.* New York: Oxford University Press.

United Nations Population Fund (UNFPA). 2001. *The State of World Population 2001—Footprints and Milestones: Population and Environmental Change.* New York: UNFPA.

U.S. Department of Energy. 2010. Critical Materials Strategy.

Weizsäcker, E., A. Lovins, and H. Lovins. 1997. *Factor Four: Doubling Wealth–Halving Resource Use.* London: Earthscan.

World Commission on Environment and Development. 1987. *Our Common Future.* A Report to the UN General Assembly. Oxford University Press.

CHAPTER 13

Toward a New Paradigm

If we want things to stay as they are, things will have to change.

—GIUSEPPE TOMASI DI LAMPEDUSA

THE LEOPARD

The rate of population is slowing, but the population continues to grow. Each new arrival brings greater needs for food, water, shelter, and durable and non-durable goods of all kinds. Meanwhile, the global economy is expanding at a rapid rate, translating to unprecedented levels of consumption, resource extraction, and associated environmental impacts. Many, including a number of industry leaders, believe that an industrial raw materials supply problem is on the horizon. Others fear sharp rise in environmental impacts from increasing extraction activity.

Such concerns are relatively new. For the past several hundred years only a handful of countries containing only a small fraction of the world's people accounted for the great majority of natural resource consumption. That the economic genie escaped from the proverbial bottle at about the same time that population numbers exploded throughout the world fundamentally changed global dynamics. Suddenly the world's economic elite, which had long enjoyed uncontested access to global minerals deposits, timber, and energy resources, found themselves in competition for materials needed to support their economies and lifestyles. That many of the leading economies had come to rely on massive net imports for key raw material supplies accentuated concerns.

The recent past characterizes the present. Complicating the current situation is awareness among citizens of less economically developed countries of the sharp disparity that exists in resource consumption between the highly developed and lesser developed economies. This is also a cause of growing discontent as seen in recent statements of intellectuals and political leaders, and in developing-country-inspired UN initiatives.

Particular attention is currently focused on future minerals supplies. Some view the minerals situation with alarm, calling for urgent attention of world governments to a goal of dematerialization in order to ensure future supplies and avoid substantial environmental degradation. Others dismiss concerns about raw material availability, pointing to a long history of problem avoidance through innovation and technology development guided by free market price signals. This latter view stems from awareness of a number of past predictions of resource shortage, none of which materialized thanks to unanticipated technological advances.

Several recent reports have acknowledged the past success and future promise of ongoing technology development, but emphasized that governments and corporations worldwide have generally been reducing R&D budgets in recent decades. Noting that problems appear to be developing in the face of current levels of expenditure and research activity, they caution that if society is to rely on technology to ensure global resource sufficiency and environmental sustainability, more rather than less investment will be needed.

Those who believe that technology development alone may not be sufficient to meet the unprecedented resource and environmental challenges of the future seek additional solutions. Among these are measures to reduce consumption in developed countries with a goal of reducing pressure on global raw material reserves and production regions. A counter view is that developed country consumption and resource importation provides income and jobs to developing countries that can serve to drive demographic transition, ultimately leading to stabilized populations and greater protection of the environment. Others point out that the high consuming countries are unlikely to reduce their consumption in any event.

Considering all of these factors, scientists engaged with the United Nations have mounted the dematerialization initiative—one that seeks steep reductions in global raw material consumption without reductions in consumption *per se*. In accordance with Factor Four thinking, it is undoubtedly possible to do more with less, short of doing without. But it will be a difficult balancing act to achieve significant gains globally, requiring action across the global community, with leadership provided by the high consuming nations. Success in achieving marked change in materials consumption will require more than the work of a few thoughtful, dedicated souls. Broad acceptance of the concept and sustained action will also be needed on the part of many within government, business and industry, the scientific community, and individual citizens.

One thing is increasingly clear: the most economically developed nations need to find ways to take greater responsibility for their consumption. Whether taking the lead in stimulating greater investment in materials-oriented innovation and technological development, changing policies to allow greater reliance on domestic raw material reserves, or actually tempering domestic consumption, bold action would appear to be needed.

Rethinking Raw Materials Strategies

The Role of Governments

Actions that may be taken to address increasingly daunting problems will be most effective if they are part of a coordinated strategy. Crafting such a strategy will require answers to a number of questions that every governmental and societal leader needs to thoughtfully consider. Key questions are:

- What are the risks of maintaining the present course and assuming that world markets and economic systems will be sufficient to take care of whatever social and resource allocation issues that may arise?
- Are there enough indicators of trouble ahead to justify attention to anticipated resource supply concerns and related environmental issues?
- Are issues of global equity something that require attention from a moral perspective? In this regard, what are the responsibilities of the developed countries? Developing countries? What is the downside risk of ignoring equity concerns?

- Is the future necessarily a win/lose proposition—a zero sum game in which some countries "win" and others "lose"? Does it need to be that way?
- Is the best global strategy for countries to largely ignore resource concerns and simply invest heavily in military preparedness so as to prevail in resource conflicts?
- Is continued adherence to growth-oriented economic models a good strategy for the future? In light of the likelihood of population stabilization is re-examination warranted?

For political leaders, the almost overwhelming pace of change presents a considerable challenge. The natural tendency of humans is to resist rather than embrace change. Political leaders are often tempted to tell people what they want to hear. Yet, for those who worry not so much about the next election, but about future generations as well, the challenge is to recognize current and emerging problem areas, to consider and evaluate the full array of options, and to convincingly articulate needs for change if that is what is needed—even if that runs counter to public preferences.

Complicating matters is climate change, as it has the potential to fundamentally change the situation with regard to water availability, food production, habitable land area, and more. What should be done about this problem? Ignore it and hope for the best? Take steps to reduce emissions? Plan for mitigation?

Technology Development

History has shown that governments are not skilled in creating new and better products at low cost for society's benefit, or in efficiently allocating scarce resources. But there is much that they can do to effect change. Governments can, as indicated earlier, play a major role in bringing about change through strong and continuing public investment in R&D, creating a level playing field for business and industry, playing a collaborative role in bringing together various sectors of society in technology development efforts, creating incentives or disincentives aimed at achieving desired change, and establishing laws and regulations.

So there is much that could be done. Within the United States, obvious priorities include:

- More aggressive pursuit of renewable energy technology and infra-structure development.
- Significantly increasing funding of industrial raw materials oriented R&D encompassing exploration, development, conversion, use, substitution, and end-of-life recovery and recycling.
- Acceleration of research focused on development of new and improved renewable materials and energy.
- Expansion of investment aimed at improving materials recovery and recycling infrastructure nationwide.
- Creation of incentives for increasing citizen participation in recycling.

Environmental Law and Policy

Governmental bodies around the world define how various segments of society must operate with regard to the environment through regulations, policies, and other mechanisms. In crafting regulations and policies, of-ficials operating in democratic governments react to input from citizens, business interests, and in the process often find that citizen and business/industry input is conflicting. Finding a balance when the views of various interests conflict is one of the hardest jobs of government.

Matters concerning environmental law are particularly tough, since ac-tions on one side of a border often result in impacts on the other side of the border. The same is true even for countries separated by hundreds or thousands of miles. But those impacted are never provided the opportunity for input, or seldom even considered. And so it is that when citizen input suggests that resources should not be extracted in order to protect the local environment, or when governments create barriers to resource procurement that have the effect of raising costs of local production, the result is often im-pact on environments in regions where citizens have little say in government policy or where environmental laws are relatively lax. This is the kind of dynamic that has led, in part, to extraordinary levels of resource importation on the part of the world's largest economies, including the United States.

Change is needed. Top priorities regarding environmental policy include:

- Reexamination of environmental policies, laws, and regulations that singly or in combination serve to prevent, delay and/or increase costs or uncertainty of domestic industrial raw materials extraction, and make changes so as to lower barriers.
- Amendment of federal laws to require assessment of likely impacts to ecosystems and associated flora and fauna in other world regions before designating domestic lands as off-limits to raw materials extraction. Apply changes to all policymaking, including changes in federal lands designations and decisions made under the Endangered Species Act (ESA).
- Modification of the Antiquities Act which allows a U.S. President to unilaterally designate vast areas of federal lands as off-limits to raw materials production.
- Modification of federal forest policies regarding management of government-owned land.

Reducing Barriers to Domestic Raw Materials Production

A 2015 examination of mining permitting and operation in the United States found that unexpected and lengthy delays in obtaining mining permits set the U.S. apart from all other developed nations. The study found that whereas it takes from seven to ten years to obtain a mining permit in the U.S., in countries with similarly stringent environmental regulations the average time to obtain a permit is two years. Report authors noted that unexpected delays in permitting processes alone can reduce a typical mining project's value by more than one-third. Further, the higher costs and increased financial risks associated with prolonged permitting can cut the expected value of a mine in half even before production begins.

Other studies have revealed a pattern of frequent litigation involving proposed timber harvesting on lands administered by the USDA-Forest Service and Department of Interior Bureau of Land Management. Impacts go far beyond the number of suits filed and the actual costs of litigation. On average, two lawsuits are filed each week under the National Environmental Policy Act, a number that does not count suits brought under the Endangered Species Act or other measures. One result of litigation is that thousands of personnel hours are spent on development of information needed in litigation defense, a diversion of effort that leads to considerable foregone and delayed work, and increased costs of land management. Additional hours are invested in anticipation of litigation. While it might be said that such planning is good in that every aspect of potential impact is anticipated before field work begins, a clear result of excessive and defensive-oriented planning is that harvesting costs are increased.

While steps have been taken in recent years to reduce frivolous litigation, a more basic problem remains largely unaddressed. Difficulty arises from multiple laws and regulations that overlap, and in some cases conflict, and for which enforcement occurs across multiple agencies. This situation exists within federal law and between laws of various states and the federal government. The effect is to greatly increase the difficulty of strictly aligning with every aspect of every law, creating entrée for litigation and raising land management costs. A concerted focus on reducing redundancy in laws and coordinating implementation is badly needed.

Changing Procedural Requirements
for Shifts in Land Designation

As things now stand, proposals for industrial expansion in the U.S. must be thoroughly examined through a formalized and exhaustive environmental impact evaluation. Curiously, no similar requirement exists for evaluation of government-initiated and environmentally-oriented actions, such as those calling for establishment or expansion of reserve areas or development of new land management rules or restrictions. For example, environmental planning in the U.S. seldom includes any consideration of the reality of consumption or of the accompanying need for raw materials, with little or no thought given to possible unintended consequences. As a result, environmentally-based decision making has routinely fostered increasing raw material importation along with a transfer of associated environmental impacts. Because the global impacts of local actions are so often overlooked, the time may have come to *require* global thinking in environmental decision making.

Changes are needed in the way that land areas are designated as parks, preserves, reserves, and other set-aside areas. Similar to the environmental impact statement that today is required of firms seeking permission for expansion of commercial activity, any proposal for land redesignation should include a comprehensive evaluation of the probable impact of such a redesignation on regional and global raw materials flows as well as associated global environmental impacts. Formal review of such proposals could serve to bring broader considerations into environmentally-inspired decision making. The U.S. could, for instance, require Global Environmental Impact Assessments of all proposed actions that would place large land areas off limits to minerals development, extraction of energy resources, or periodic forest harvesting. Such a requirement would help to ensure that actions vis-à-vis environmental sustainability weren't haphazard, but were instead based on consideration of all the essential pillars of sustainability—ecological, economic, and social.

Regarding the Antiquities Act, the measure that allows U.S. Presidents to set aside vast areas without discussion or debate, sometimes for purely political reasons, it is time for sharp modification or repeal. As passed by Congress in 1906, the intent of the Act was to allow a President to declare, at his or her discretion, "historic landmarks, historic and pre-historic structures, and other objects of historic or scientific interest that are situated upon the lands owned or controlled by the Government of the United States to be national monuments." The law further provided that in making such designation, the President was to confine such designations "to the smallest area compatible with proper care and management of the objects to be protected." Recent use of this act by the chief executives of both major parties has demonstrated less and less restraint in use of the law.

Examples of questions to be addressed in a Global Environmental Impact Assessment (again using the U.S. as an example of how this might apply) would include:

- Is the proposal for a change in management based on thorough scientific evaluation?
- How will adoption of the proposal impact net imports for fuel and non-fuel minerals, wood and wood products, and other materials?
- Will adoption of the proposal be likely to shift raw materials extraction and environmental impacts elsewhere?
- Are shifts in extraction activity likely to be to other regions within the U.S. or to another country?
- Are regions to which extraction activity is likely to shift characterized by scenic beauty, sensitive watersheds and ecosystems, or populations of rare and endangered species?
- If a proposal will result in reduced production or extraction of a particular raw material, is there an accompanying initiative to reduce consumption of that material through greater efficiency in use, recovery and recycling, or substitution?

Thinking in a different way could lead to a new and more responsible approach to environmentally-oriented decision making. For instance, if a society has domestic reserves of a particular resource, but concludes that the environmental impacts associated with extraction and processing are unacceptable, then perhaps that society and leadership might become more serious about recovery and recycling of high impact materials or materials substitution. As long as impacts of consumption remain out of sight and out of mind it is highly unlikely that any groundswell of public opinion against unrestrained consumption will develop. Conversely, if the public is forced to deal to a greater extent with issues related to resource extraction, processing, and waste disposal, then it is far more likely that serious discussion about consumption will reach the political agenda. Were such thinking institutionalized, society might even begin to seriously address consumption itself.

Some argue that reducing procurement of raw materials from developing nations would also reduce payments to those nations, with negative impacts on income. However, it is difficult to convincingly argue that raw material importation and basic processing of raw materials is part of a strategy to help developing nations grow economically, since a potentially more effective and far less invasive policy would be to focus on a transfer of clean industry and service jobs to them.

Modifying Federal Forest Policies Regarding Management of Government-Owned Land

Forests managed by federal agencies, and particularly the U.S. Forest Service, are in a state of neglect, a lasting legacy of failures in federal land management policies. Combustible fuels are accumulating rapidly with a large gap between annual growth rates and annual removals, high and rising natural mortality, and increasing incidence of catastrophic fire. This situation exists despite the preponderance of softwoods in the nation's federal forests and continued reliance on Canada for about a third of annual softwood lumber consumption.

Because failure to manage lands to provide a sustainable flow of goods and services more closely aligned to production potential is creating environmental and social impacts outside the borders of the United States, a greater focus on land stewardship in federal land mangement is needed. Continuing reliance on Canada for wood causes other countries, which would otherwise obtain wood supplies from Canada, to seek sources elsewhere, resulting in a cascade of global impacts.

One aspect of needed change is a shift away from the currently-held idea that trees can only be harvested as part of thinning projects, and then only for the purposes of reducing fire risks or spread of disease. Managing and periodically harvesting trees to fulfill diverse economic and ecological goals is a legitimate activity, including within lands managed by the U.S. Forest Service. National forests were specifically established for the purpose of protecting watersheds and providing a continuous supply of timber, delineating a clear difference between national forests and national parks.

That more could be done within high-consuming societies to focus attention on consumption and on strategies for reducing raw material depletion and global environmental impacts is obvious. That there is a need to do so is likewise increasingly apparent. Whether there will be the necessary political will at any point in the foreseeable future remains to be seen.

The Role of Business and Industry

It is tempting to conclude that only government institutions are equipped to take on the complex issues the world is facing and the multifaceted solutions that will be required. Yet, if change is to occur on a scale sufficient to accommodate the numbers that are coming, and without calamitous outcomes for humanity and the environment, then responsibility for change will need to be shared throughout society. The private sector, and major players within it, will clearly need to be at the forefront of effecting change, innovating, and bringing new discoveries into reality.

Increasing Materials Use Efficiency

Because gains from research and development often benefit society beyond the realm of an individual company or industry, both the costs and actual conduct of research are often shared by industries and governments. Materials-oriented R&D in the United States and around the world, for instance, is conducted by both private sector companies and national governments and by universities with private sector or government funding. In the U.S., materials-oriented research on the part of the federal government is being coordinated by the Department of Energy, with a focus on extending non-renewable materials supplies (and especially rare earth metals) through increases in materials use efficiency, identification and/or development of substitute materials, recycling, and reducing environmental impacts of mining. The agency has also mounted joint efforts in cooperation with industry. A number of other developed nations have mounted similar initiatives.

Such efforts routinely lead to increasingly lighter and less resource intensive products, development of substitute materials when preferred raw materials become scarce or too expensive, engineering of entirely new materials and technologies that lead to lighter, lower impact products, and more. Despite market-driven and ongoing technology improvement, consumption of virgin materials continues to grow rapidly. More rapid progress is needed than is being realized at present. The challenge is illustrated by the determination that limiting the increase in global materials consumption to, for example, 40 percent above consumption at the beginning of the 21st century will require developed nations to reduce materials use by three to fivefold and require moderation in materials use by developing nations as well.

Japan, which is among the most aggressive of nations with regard to investment in R&D, is perhaps a model for what needs to happen more broadly around the world. In 2000, Japan's Diet passed into law the *Basic Act for Establishing a Sound Material-Cycle Society.* The measure defined a sound material-cycle society as one in which "the consumption of natural resources will be conserved and the environmental load will be reduced to the greatest extent possible, by preventing or reducing the generation of wastes, . . . by promoting proper cyclical use of products, . . . and by ensuring proper disposal of circulative resources not put into cyclical use."

The Act also clarified the responsibilities of the central and local governments, business operators and citizens, and established a blueprint for future action, including the establishment of a basic strategy for moving forward. Japan today has in place a suite of laws that promote waste minimization and effective utilization of resources. Core elements of the strategy include preventative laws and measures, funding and implementation of research and development, and education and engagement of the public. The ultimate goal is a closed-loop society.

Reducing Impacts of Materials Procurement and Use

Until quite recently, the world was a mysterious place. The Far East was, well, far away, as were the jungles of the Amazon and the diamond mines of South Africa. So, when raw materials or products from such places found their way into world markets, most people in the distribution chain knew little or nothing about the specifics of their origin and had little opportunity to find out. However, global communication and air travel have brought the far reaches of the world closer for ordinary citizens in developed and developing countries alike. Information is only a click of the mouse, a flick of the remote, a plane ticket, or a call to a raw material or product certifier away. These developments are affecting our raw material sourcing.

Manufacturers can no longer plead ignorance about the origin of raw materials. Distributors, large volume buyers, and individual consumers alike expect top quality at competitive prices, but they also increasingly expect verifiable assurances that neither environmental damage nor human exploitation is linked to their purchases. Consequently, players throughout product supply chains must use due diligence in determining the real impacts of producing, distributing, using, and disposing of the goods they produce or distribute. Now another imperative is emerging—that raw material inputs, and in particular virgin raw material inputs, will be substantially reduced in manufactured products.

Business and industry are vital partners in changing the trajectory of resource consumption. Keeping in mind that there is no magic bullet, and that rapid progress is needed on a number of fronts simultaneously, companies must increasingly pay attention to all segments of their supply chains. With regard to materials use efficiency, not only what a company consumes, but also what it wastes, are important considerations. What needs to be known about waste products are the quantity and character of waste, and how much is spent to handle, store, and dispose of it; how much waste is recycled either internally or externally; how much more could be; and whether markets might exist for some proportion of the current waste stream. Ongoing evaluation of product designs is also important, with an eye toward improvement in materials use efficiency and recovery of components and materials at the end of product life.

To effectively address environmental issues, a business enterprise needs to understand what the impacts of its operations are throughout its supply chain, including those associated with resource procurement and processing. The source of products must be well understood—where raw materials come from, what is known about environmental impacts of gathering and processing these raw materials, whether products are veri-fiably of legal origin, and so on. Knowledge regarding energy use at each point in production systems is also important so that industries can quan-tify where the greatest energy reduction could be achieved, thereby estab-lishing priority targets for strategic investment. In general, all but several of the major categories of measurable environmental impact are directly or indirectly associated with energy use. Thus, even without conducting a full assessment of operations, by focusing on reducing energy use, most business enterprises can markedly reduce environmental impacts.

Another way to reduce environmental impacts is to improve the useful lives of products, either by designing to increase product durability or extending support for maintenance of products to allow longer service life. Using the electronics sector as an example, the U.S. Environmental Protection Agency has estimated that extending the useful life of personal computers from four to six years would annually reduce resource con-sumption by 439,000 tons, carbon dioxide-equivalent emissions by over 24 million tons, and waste disposal in landfills by 316,000 tons. Similar gains could be obtained through production of longer-lived cell phones, tablets, and other electronic devices. Because this approach would likely lead to reduced sales volume, product durability has generally received less attention than materials use efficiency.

A commitment to operating with environmental concerns front and center is not trivial. Over any long period of time, success in maximizing environmental performance requires buy-in at all levels of a business entity, from the chief executive, to product designers, to line supervisors and employees, and even the custodial staff. Consideration of environ-mental impacts becomes, in effect, a part of the same discussions that may today be focused on new product development or profitability alone.

Standards are now in place for reporting environmental and social per-formance, the best known of which are those developed by the Global Reporting Initiative. This organization was created by two U.S. non-profits: the Coalition for Environmentally Responsible Economies and the Tellus Institute. The Global Reporting Initiative has developed a set of principles for responsible environmental, social, economic, and governance conduct and transparent reporting of performance in each of these areas. The first version was issued in 2000. As yet, there are no similar initiatives designed to track progress toward reduced raw materials consumption.

In the early part of the 21st century investor groups began to lean on industries and specific companies to improve environmental performance. In the fall of 2006, a group of major institutional investors called on American companies to follow their European counterparts in using the Global Reporting Initiative to improve their public disclosure to shareholders on pressing environmental and social issues. Citing a vast reporting gap between U.S. and overseas companies, nine investors, representing more than $300 billion in assets, sent letters to S&P 500 companies requesting GRI participation. As of early 2015, 73 percent of these companies reported performance through the Global Reporting Initiative. As of early 2017, 92 percent of the 250 largest corporations globally were reporting environmental performance through the GRI.

In 2011 a coalition of institutional investors, through what is called the Carbon Disclosure Project, also began asking corporations for information about their carbon emissions and plans for cutting them. The stated purpose was to assess how well companies would be likely to do in a carbon-constrained economy, thereby allowing informed determination of risk in investment portfolios. In early 2015, 767 investment organizations, representing over $92 trillion in assets, were involved. Over 5,000 firms were engaged in carbon reporting, including 81 percent of the world's 500 largest corporations.

Another example of shareholder pressure on the private sector vis-à-vis environmental performance is the launch, in late October 2013, of a coordinated effort to spur 45 of the world's top oil and gas, coal and electric power companies to assess the financial risks that climate change pose to their business plans. A group of 70 global investors managing more than $3 trillion of collective assets, most of them based in the U.S. and Europe, sent letters to the fossil fuel companies requesting detailed responses before their annual shareholder meetings in early 2014. This effort is the first of its kind, and appears to signal a new relationship between society (as represented by the investment community) and business and industry.

In addition, a new concept—the Benefit Corporation (B-Corp)—has been introduced. B-Corps organizations must pledge, as part of their corporate structure, to operate in a socially and environmentally responsible manner. Recognized in early 2017 by 30 states and the District of Columbia within the U.S., B-Corps are given greater latitude in pursuing environmental and social responsibility than those of other corporations that are required by law to prioritize the financial interests of shareholders over the interests of workers, communities and the environment. As of February 2017, there were over 2,000 B-Corporations in 50 countries worldwide, with the majority of these based in the United States and Canada. Further, governments around the world have initiated environmentally preferable purchasing programs.

Fortunately, there are resources for companies striving to improve materials use efficiency and overall environmental performance. For example, all of the leading accounting firms now offer consulting and analytical services on measuring and improving environmental and social performance. And, through a 2014 initiative of TruCost—an environmental consulting firm—and the GreenBiz group, companies are provided information about the scale of their environmental impacts, formatted to allow direct comparison with financial metrics. Called the Natural Capital Leaders Index, information is generated through use of environmental impact data disclosed by companies to produce a complete environmental footprint covering that companies' own operations and related supply chains. One aspect of the program is a public listing of companies that are at the forefront of decoupling revenue from environmental impact. Similar programs are offered by other companies.

The Role of Individual Consumers

The importance of an informed citizenry cannot be overemphasized. There is much that an individual can do to contribute to long-term solutions, including:

- Becoming as informed about environmental issues as possible.
- Reading widely. Being receptive to new ideas. Searching out opinions representing all sides of an issue, and avoiding labeling of those who may disagree.

- Taking steps to learn more about the world beyond one's country of residence.
- Avoiding being made a victim of misinformation by questioning everything read or heard. Making every effort to find out what is true and what is not.
- Challenging misinformation wherever it occurs, and taking steps to correct the record at every opportunity.
- Thinking about global implications of consumption and of local, regional, and national environmental policies and decisions.
- Seeking opportunities for informed, rational discussion about population, consumption, and global equity issues.
- Paying particular attention to what children are being taught about environmental issues and encouraging the teaching of systematic thinking. Young people will assume decision making roles in a remarkably short time, and what they learn or fail to learn at an early age will profoundly influence their thinking.
- Encouraging government, business, academic, and societal leaders who are working to bring about positive change.
- Participating fully in community recycling programs.
- Taking steps to reduce personal consumption or consumption impacts.
- Considering carefully before purchasing: do I really need this?
- Paying attention to environmental product labels. Consistently selecting low energy consuming, low embodied energy products when making purchase decisions.
- Avoiding lowest price vendors, instead basing purchase decisions on quality, durability, and longevity of goods purchased.
- Using products made of sustainably produced renewable resources rather than non-renewable resources at every opportunity.
- Taking the time to understand exactly what any organization seeking donations stands for, and what it does, before providing support. If activity is found to be primarily centered on demonizing industry and raw material extraction, then further consideration of where to place conservation contributions is probably warranted.
- Seeking to become part of the solution and not part of the problem.

Confronting Population Growth

Stabilizing Population Numbers

Growth of human numbers underlies virtually all of the problems that society is facing today, including those linked to consumption. It continues to be ignored as part of potential solutions.

That population growth will halt within the next century or so is considered a near certainty, but the level to which it grows in the meantime matters. Whereas discussion tends to focus on the medium projections of growth—9.0 to 11.2 billion by 2100—the range of projections—8.9 to 13.3 billion—is wide.

If, before stabilizing, the population continues to grow to 12–13 billion rather than 9–11 billion, meeting human needs and demands while protecting the Earth's environment will be significantly more challenging than if stabilization occurs sooner. Population growth remains an issue, and there is much that could be done, that does not involve abortion, to ensure population stabilization at a lower rather than higher level.

Many organizations have developed lists, based on studies of population dynamics, of steps that could be taken to slow growth rates and stabilize population. Among the most comprehensive of these is the following list from the Worldwatch Institute:

- Provide universal access to safe and effective contraceptive options for both sexes.
- Guarantee education through secondary school for all, especially girls.
- Eradicate gender bias from law, economic opportunity, health, and culture.
- Offer age-appropriate sexuality education for all students.
- End all policies that reward parents financially based on the number of children they have.
- Integrate lessons on population, environment, and development into school curricula at multiple levels.
- Put prices on environmental costs and impacts to allow couples to quantify the cost of an additional family member by calculating taxes and increased food and other costs to facilitate individual family planning.

- Adjust to an aging population instead of boosting childbearing through government incentives and programs.
- Seek to convince leaders to commit to stabilizing population growth through the exercise of human rights and human development.

These actions are based on recognition that people, and especially women, who receive formal education are more likely to delay initiation of families and to have fewer children than those who do not have the benefit of education. Similarly, women who are empowered in society in areas of law, economic opportunity, health, and culture equal to men are likely to have children later in life and fewer children than those deprived of social equality.

These things have been known for some time, and there are a number of initiatives underway around the world based on knowledge of such strategies. But there is much more that could be done.

Proactive Planning for Population Stabilty

The flip side of the population equation is solving ahead of time potential problems linked to a no-growth society. At the point at which population does stabilize (which it must), then society will need the benefit of forward-looking thinking and planning for a no-growth society. In this case much of the focus will need to be on growth-oriented economic systems. While the global economy has for decades grown more rapidly than population, stable populations and increasingly convergent standards of living are likely to significantly dampen consumption growth at the point that growth in numbers reaches a peak.

Regardless of the level to which population rises, or the timing of stabilization, the voices now expressing concern over slowing of population growth can be expected to become increasingly shrill as the growth rate continues to slow. Proposals for incentives to renew population growth are possible, and even likely. In fact, such ideas are already being floated. But it is critically important that the nations of the world proactively learn how to deal with zero or even negative population growth, and to hold steadfast against any proposals for incentivizing a return to a growth policy.

Closing Thoughts

Growth is a central tenant of thinking within business and industry, and within the halls of government around the world. Companies must grow to remain competitive. Economies must grow to remain healthy. Growing populations are needed to support social programs for the elderly. And so on. But it is physically impossible for populations to grow forever in a closed system. And it is unlikely that consumption or economies will grow much beyond the point at which population growth ceases. Without any doubt fundamental changes lie ahead, and advance thinking and planning are needed to allow a smooth transition to the new reality—one in which nations will seek to pursue prosperity in the absence of growth.

For the rest of this century, and perhaps longer, the population will grow, as will competition for the world's resources. How to provide for wants and needs while also protecting the global environment presents a non-trivial challenge. Continued technology development may represent the closest thing to a silver bullet solution to the problems society faces. However, indications are that success will require investment well beyond current levels. Addressing social equity issues is likely to be more difficult, in part because free market economics invariably provide greater leverage to those with the most financial resources.

The less-well-off in terms of material possessions are steadily gaining access to a larger piece of the pie, while those at the top of the economic ladder are doing everything they can to remain in a dominant position. It is a process that is certain to continue. Often caught in the middle are the environment and the world's natural systems, and those who live in regions from which raw materials are obtained. But the playing field isn't level, with some nations much better positioned to protect their environments than others, and therein lies a second significant problem.

The economically most advanced countries are able to both consume the greatest quantity of goods, and invest the most in environmental protection. It is a formidable combination that allows citizens of those countries to afford high consumption as well as a high degree of protection from environmental insult. Consumption requires the use of raw materials, some of which trigger significantly lower environmental impacts than others, but environmental impacts nonetheless. Those raw materials have to come from somewhere. So when citizens of the most economically developed countries allow or encourage environmental policies that have the effect of blocking domestic production of raw materials, the effects are to directly impact environments elsewhere.

It doesn't matter if citizens feign ignorance, act as they do not in a calculated way but by default, or continue present practices because they simply don't care, the fact of the matter is that people and environments all over the world are impacted by first-world consumption, and often negatively. It is also likely that consumption practices within the most highly-developed countries are narrowing the options of future generations.

It would be an injustice to suggest that the United States and other highly developed economies have done nothing to address the issues herein. The past several decades have seen introduction of forest certification, international attention to tropical deforestation, growing awareness and use of environmental life cycle assessment, and involvement in leading corporations in organizations such as the Global Reporting Initiative, Coalition for Environmentally Responsible Economies, and the Carbon Disclosure Product. These are important and meaningful developments. But the clock is ticking. The long standing obsession with growth continues, and the firmly established practice of vigorously objecting to domestic raw material production while minimizing consideration of raw material consumption shows little indication of change.

We need to change what we are doing: to develop greater awareness of the local and global impacts of our consumption, learn how to distinguish high impact products from less impacting products, consume smarter, and accept more responsibility as a nation for those things that we do consume. We need to think about issues of global equity and consider how we might view the situation were the shoe on the other foot. We also need to think carefully about what characteristics or actions should define an environmentally-concerned citizen, leader, or environmentalist, and withhold support from those who seek to maintain the current state of affairs. And we need to carefully think about population policies and the implications of both further growth and eventual population stabilization.

From an ethical perspective, the need for global thinking is obvious. Contrary to common practice today, high consuming nations need to much more often consider the question "Why *not* in my backyard?"

References

Brainerd, E. 2014. "Can Government Policies Reverse Undesirable Declines in Fertility?" *World of Labor,* May.

Engelman, R. 2012. "Nine Population Strategies for Stopping Short of Nine Billion." In *State of the World 2012: Moving Toward Sustainable Prosperity,* Chapter 9. Washington D.C.: Worldwatch Institute.

Fischer-Kowalski, M., M. Swilling, E. von Weizsäcker, Y. Ren, Y. Moriguchi, W. Crane, F. Krausmann, N. Eisenmenger, S. Giljum, P. Hennicke, P. Romero Lankao, A. Siriban Manalang, and S. Sewerin. 2011. *Decoupling Natural Resource Use and Environmental Impacts from Economic Growth- A Report of the Working Group on Decoupling to the International Resource Panel.* United Nations Environment Program (UNEP).

Global Reporting Initiative (GRI). 2015. "GRI Guidelines and Standards Setting," Accessed December 2, 2015.

Government of Japan. 2000. *The Basic Act for Establishing a Sound Material-Cycle Society: Act No. 110 of 2000.* Ministry of the Environment.

Government of Japan. 2010. *Establishing a Sound Material-Cycle Society: Milestone Toward a Sound Material-Cycle Society Through Changes in Business and Life Styles.* Ministry of the Environment.

GreenBiz. 2014. "The Natural Capital Leader's Index, 2014," Accessed August 19, 2015.

Longman, P. 2004. *The Empty Cradle: How Falling Birthrates Threaten World Prosperity and What to Do About It.* New York: Basic Books.

Männik, K., K. Eljas-Taal, J. Angelis, and A. Rozeik. 2012. *Funding Research and Innovation in the EU and Beyond: Trends during 2010–2012.* European Commission, December.

Metzger, E., Y. Dagnet, S. Putt del Pino, J. Morgan, L. Karbassi, H. Huusko, F. Silveira, M. VanVoore, S. Haeussling, B. Watson, T. Carnac, A. Pineda, L. Pierce, A. Kelly, J. Walker, and D. Ryan. 2013. *Guide for Responsible Corporate Engagement in Climate Policy.* World Resources Institute, Caring for Climate Initiative.

Miner, A., R. Malmsheimer, D., D. Keele, and M. Mortimer. 2010. "Twenty Years of Forest Service National Environmental Policy Act Litigation." *Environmental Practice* 12(2): 116–126.

Morgan, T. and J. Baldridge. 2015. *Understanding Costs and Other Impacts of Litigation of Forest Service Projects: A Region One Case Study.* University of Montana, Bureau of Business and Economic Research, May 5.

Mortimer, M. and R. Malmsheimer. 2011. "The Equal Access to Justice Act and US Forest Service Land Management: Incentives to Litigate?" *Journal of Forestry* 109 (6): 352–358

Onorato, D. 2001. "Japanese Recycling Law Takes Effect." *Waste 360,* June 1.

PriceWaterhouseCoopers (PwC). 2013. *Measuring and Managing Total Impact—Strengthening Business Decisions for Business Leaders.*

SNL Metals and Mining. 2015. *Permitting, Economic Value and Mining in the United States.* Report prepared for the National Mining Association.

U.S. Congress. 1906. American Antiquities Act of 1906. 16 USC 431–433.

U.S. Environmental Protection Agency. 2009. *Opportunities to Reduce Greenhouse Gas Emissions through Materials and Land Management Practices.* Office of Solid Waste and Emergency Response. Sept.

U.S. Government Accounting Office. 2010. *Forest Service. Information on Appeals, Objections, and Litigation Involving Fuel Reduction Activities, Fiscal Years 2006–2008.*

U.S. Government Accounting Office. 2014. *National Environmental Policy Act: Little Information Exists on NEPA Analyses.*

Worstall, T. 2014. "Bill Gates Points to the Best Tax System, the Progressive Consumption Tax." *Forbes,* March, 18.

Epilogue

And to those nations like ours that enjoy relative plenty, we say we can
no longer afford indifference to suffering outside our borders;
nor can we consume the world's resources without regard to effect.
For the world has changed, and we must change with it.

—BARACK OBAMA

When he returned home, Daniel had experienced an unimaginable surprise. Lord Jardel, upon hearing the insightful report Daniel delivered,
arranged for a retelling directly to the King. The King, in turn, had also
been impressed by the detailed observations of Paraíso, and by Daniel's
adventurous spirit and interest in the world beyond. To Daniel's astonish-
ment, he had been appointed as assistant to the King's chief emissary, a
development that gave him a lifetime of opportunity to travel, a modest
income, and quarters within the castle walls. And, for a time, Daniel's
new status brought notoriety, most importantly among the young ladies
of the court.

During the three decades that Daniel had served his kingdom he had
visited, more than once, all of the surrounding kingdoms and fiefdoms, as
well as some that were many months distant. Conditions in most of the
kingdoms that he visited were more or less similar to his own, and only a
few came close to matching the wonders of Paraíso. As to Paraíso itself,
he had heard rumblings in his later years that problems within the magical
kingdom were beginning to surface—problems that seemed to reinforce
his earlier questions about the long-term adequacy of the royal treasury.

Over the following generations, things had gone badly for Paraíso.
Accustomed to obtaining what they wanted from surrounding kingdoms,

they had failed to take seriously forecasts of increasing competition for resources, and of rising environmental impacts linked to consumption. Some had assumed that military power would solve whatever problems might arise, but growing economic and military power in rival kingdoms had increased demands on the royal army, making it challenging and prohibitively expensive to finance far-flung operations. At the same time, the number of Paraísoans had grown rapidly, taxing the ability of the realm to provide for their needs and whims, which in turn led to growing discontent among the King's subjects. One of the things that had become quite difficult was maintaining the storyland appearance of Paraíso, and as this became more and more evident, discontent was magnified. To make matters worse, the flow of basic resources from surrounding kingdoms had begun to ebb as these kingdoms consumed more and more resources to support their own growing economies.

Eventually, simmering conflicts became resource wars which resulted in needless loss and suffering in kingdoms everywhere. Paraíso was impacted more than most, and it emerged from the chaos as a weakened state. Its leaders recognized in retrospect that they should have paid more attention to calls for greater equity among the various kingdoms. Many centuries would pass before Paraíso regained a semblance of its former allure.

Whereas peasants and nobles throughout the world had once known of Paraíso and dreamed of going there, students throughout the world later studied this once magical kingdom to examine what had gone wrong. In classroom after classroom, it was concluded in hindsight that if only a few fundamental decisions had been made differently, Paraíso would still be the jewel of the world.

About the Author

Jim L. Bowyer is Professor Emeritus, University of Minnesota Department of Bioproducts and Biosystems Engineering, and an Elected Fellow of the International Academy of Wood Science. He is also President of Bowyer & Associates, Inc., a consulting firm focused on helping organizations to improve their environmental performance, and Director of the Responsible Materials Program of Dovetail Partners, Inc., Minneapolis, Minnesota, a nonprofit organization that collaborates to develop unique concepts, systems, programs, and models to foster sustainable forestry and catalyze responsible trade and consumption.

Bowyer graduated from Oklahoma State University with a B.S. in forestry in 1964. Thereafter he earned an MS degree in forest products and wood science at Michigan State University and, after a three-year hiatus for military service in Southeast Asia, completed a PhD in wood science and technology at the University of Minnesota. He has worked as a forest aide, firefighter, timber cruiser, and wilderness area trail patrol in national forests of Wyoming and Colorado, devoted over 36 years in academic teaching and research at the University of Minnesota focused on biomaterials science, life cycle assessment, and global raw material trends, and for a decade has worked as an environmental consultant in his own company and with Dovetail Partners.

Dr. Bowyer has served as President of the Forest Products Society (1993–94) and of the Society of Wood Science and Technology (1987–88), Vice President of the Consortium for Research on Renewable Industrial Materials (1992–2003), an association of fourteen major North American Universities organized for the purpose of conducting life cycle assessments of renewable materials and products, and Board Member

(1994–2008) and Chairman (2006–2008) of the Tropical Forest Foundation, an organization with major operations in Brazil, Guyana, Indonesia, and Gabon. He was Head of the University of Minnesota's Department of Wood & Paper Science from 1984 to 1994, and Founder and Director of the Forest Products Management Development Institute at the University of Minnesota (an organization dedicated to education and development of industry professionals) from 1994–2003. He currently serves on a number of editorial boards and national committees with a focus on carbon issues and forest carbon dynamics.

He is co-author of *Forest Products and Wood Science, 1st through 5th editions,* and author of six book chapters and over 400 scientific articles. He has published widely on the topics of wood and fiber science, bioenergy, life cycle assessment, green building standards, global raw material trends, and environmental policy and is a frequent speaker nationally and internationally on these topics.

Questions for Discussion

In the hope that at least some of the issues discussed herein have stimulated interest in further exploration, the following questions are provided to facilitate thinking and discussion. While there are no right or wrong answers to these questions, there is very much a need for thoughtful consideration of the issues posed.

Chapter 1—A Conundrum

Referring to the end of Chapter 1 (page 18 and reproduced below), is there any solution to the dilemma posed? In other words, is it possible to protect the economy and environment for future generations while also acting more responsibly as a member of the global community?

- If the current strategy is maintained without change, despite the moral shortcomings of that strategy, is our country likely to be able to continue to successfully compete for the raw materials needed to support the domestic economy and current lifestyles?

- If consumption levels are voluntarily reduced, won't that ruin the domestic economy, making environmental protection less affordable and therefore less possible?

- If consumption is not reduced, but steps are taken to bear more of the environmental consequences of that consumption, won't that ruin the domestic environment outright?

Chapter 2—An Irresponsible Journey Toward Paradise

- Isn't it reasonable, and even helpful, to ship wastes to those who seek new sources of income, even if that waste is hazardous to health?

- Precisely because a number of countries aren't carefully looking out for their environments to the extent that most developed countries are, doesn't this provide an imperative for the most economically developed countries to take extraordinary steps to protect their own natural capital no matter what the consequences elsewhere?

Chapter 3—Misconceptions

- Do any examples of decision-making based on misinformation and misconceptions come to mind (for example in personal life, health care, or politics)? Are there any instances in which those involved benefitted?

- Is exaggeration or lying in support of a good cause justifiable? A good idea?

- Because scientific knowledge by its very nature changes over time, doesn't that make science an unreliable source of information in decision-making?

Chapter 4—The Power of Compounding

- Has consideration of the long-term effects of compounding in any way changed your thinking about the concept of sustainable growth?

Chapter 5—Growth of the Masses

- If the rate of population increase in the United States were to slow down in coming years, what should be our reaction? Should we adapt to slowing of growth and perhaps population stabilization, or should actions be taken to reinvigorate growth?

- If population growth is, in fact, essential for the health of our economy as some assert, what will happen when the population stops growing, something that must occur in the not-too-distant future?

- Is replenishment of younger workers through higher birth rates or immigration, in fact, crucial for a nation's ability to provide needed support for a growing older population? If it is, at what point, if ever, does the need for such replenishment end?

Chapter 6—Economic Miracles

- As emerging economies around the world continue to grow more rapidly than those of established economic powers, how are relationships of countries likely to change?

- Is continued adherence to growth-oriented economic models a good strategy for the future? In light of the likelihood of population stabilization in the not-too-distant future is re-examination warranted?

Chapter 7—Minerals and Metals Basics

- In view of high and rising minerals net imports as a percent of consumption by most of the advanced economies, and increasing global concerns about environment and resource distribution, should investment in reducing virgin raw materials consumption be a priority?

- Considering the options identified for reducing minerals consumption (pages 124–127), can you think of other options?

Chapter 8—Bio-resources and Products

- Greater reliance on wood and other renewable bio-resources could help to reduce consumption of non-renewable resources—metals, non-metallic minerals, and petroleum-derived plastics. But that will likely mean, among other things, an increase in periodic harvesting within forests, or establishment of new tree and fiber plantations, or both. What is your reaction to the prospect of:

- Increased forest harvest activity within domestic forests?

- Increased establishment of tree plantations domestically or globally?

- Increased conversion of lands to high intensity fiber production?

- If your reaction to all of these things is negative, and considering that increased recycling is only part of the answer, are there other alternatives that would help society to shift away from non-renewables?

- How is it possible that greater use of wood by society can help to retain forests for future generations?

Chapter 9—The Energy Juggernaut

Energy consumption, and especially consumption of fossil fuels, is a major contributor to a number of adverse environmental impacts.

- Should reduction of fossil fuel consumption be a high priority?

- If yes, what measures, if any, would you support to bring about reduced consumption?

- A shift in taxes from income to fossil energy (i.e. high gasoline taxes, other taxes on fossil fuels)?

- A carbon tax?

- Tax incentives or other measures to encourage greater building energy efficiency (beyond what is being done already)?

- Subsidization of renewable energy to encourage alternative energy innovation and infrastructure development?

- Other?

Chapter 10—Seeking Silver Bullets

- Other than the examples given in this chapter, can you think of other products or practices that are promoted as simple solutions to difficult-to-solve problems?

- Why is systematic or life-cycle thinking essential in evaluating proposed solutions to complex problems?

Chapter 11—An Environmental President

- Should an environmental leader be expected to realize that actions taken to protect the environment in one region can have negative and often profound impacts on the environment of other regions, and to act accordingly?

- A national leader who might propose steps to encourage greater raw materials procurement domestically will almost certainly be labeled "anti-environment" within some quarters. Would you be inclined to give serious consideration to what was being proposed, or to join in opposition to any such change?

- Should standards for judging environmental leaders be broadened to include demonstration, in words and actions, of a consistent world view when advocating domestic environmental initiatives?

Chapter 12—Rethinking Consumption

- If you were in a high leadership position and responsible for substantially reducing virgin industrial raw materials consumption within your country, where would you start?

- Continuing with the question posed above, how would you gain the public's support for what you were attempting to do?

- If a primary strategy for increasing the efficiency of raw materials use were reliance on technology—something that by any measure continues to be phenomenally successful in bringing solutions to problems of all kinds—what do you think might be done to sharply increase the rate of innovation and development?

Chapter 13—Toward a New Paradigm

Several of the questions below are from the list of considerations identified as critical for any national leader.

- Is the future necessarily a win/lose proposition? A zero sum game in which some countries "win" and others "lose"? Does it need to be that way?

- Is the best global strategy for countries to largely ignore resource concerns and simply invest heavily in military preparedness so as to prevail in resource conflicts?

- Should issues of global equity be given consideration in national decision-making? In this regard, what are the responsibilities of the developed countries? Developing countries? What is the downside risk of ignoring equity concerns?

- A requirement that proposed land re-designations be preceded by completion of a Global Environmental Impact Assessment (GEIA) might result in greater introspection on the part of political leaders before action is taken. But that could lead to greater resource extraction domestically. Would you favor establishment of a GEIA requirement or something similar?

- Considering the list of things that individual citizens could do to help change course (pages 254–255), which three do you believe are the most important? Any things that should be on the list which aren't?

Index

Introduction to Forest Ecology and Silviculture

Second Edition

By
Thom J. McEvoy
Associate Professor and Extension Forester
University of Vermont

Natural Resource, Agriculture, and Engineering Service (NRAES)
Cooperative Extension
152 Riley-Robb Hall
Ithaca, New York 14853-5701

NRAES–126
September 2000

ISBN 0-935817-55-7

Library of Congress Cataloging-in-Publication Data

McEvoy, Thomas J. (Thomas James), 1953-
    Introduction to forest ecology and silviculture / by Thom J. McEvoy.--2nd ed.
        p. cm. -- (NRAES ; 126)
    Includes bibliographical references (p.).
    ISBN 0-935817-55-7 (pb : alk. paper)
        1. Forests and forestry. 2. Forest ecology. 3. Forests and forestry--Northeastern States.
4. Forest ecology--Northeastern States. I. Natural Resource, Agriculture, and
Engineering Service. Cooperative Extension. II. Title. III. NRAES (Series) ; 126.

SD391 .M52 2000
634.9'0974--dc21

                                                    00-030490

Natural Resource, Agriculture, and Engineering Service (NRAES)
Cooperative Extension • 152 Riley-Robb Hall
Ithaca, NY 14853-5701
Phone: (607) 255-7654 • Fax: (607) 254-8770
E-Mail: NRAES@CORNELL.EDU • Web site: WWW.NRAES.ORG

Table of Contents

List of Figures

List of Tables and Sidebar

Preface to the Second Edition

In 1987 I received a small grant from the University of Vermont Extension Service through its Renewable Resources Extension Program to develop, deliver, and evaluate an educational program for loggers in Vermont. The primary subjects of the program were forest ecology and silviculture. Over the course of the succeeding six years, the three-day curriculum — known as the Silviculture Education for Loggers Project — was delivered three times throughout the state. About 250 loggers completed the program, nearly half the population of full-time loggers in Vermont. The program was so successful that it served as a model for a national pilot program known as LEAP — Logger Education to Advance Professionalism. From 1991 to 1994 the LEAP Program was piloted in thirteen states representing every major timber type in the U.S. Now LEAP — although greatly expanded beyond its original ecological focus — is serving as a national model at a time when the forest industry has identified logger education as a primary thrust of its "Sustainable Forestry Initiative."

During the original Silviculture Education for Loggers Project we developed two booklets for workshop participants. The first, called "Forest Ecology for Loggers," coauthored with Yuriy Bihun, was published in 1990. The second, called "Silviculture Handbook for Loggers," which relied heavily on knowledge from the forest ecology booklet, was published in 1993. Both publications were very well received by a wider-than-intended audience.

In 1994 Brian Stone, Chief of Forest Management, Vermont Department of Forests, Parks, and Recreation, asked if it was possible to combine the two publications in such a way as to broaden the audience to include woodland owners, since many forest owners had already found the original booklets useful. As is usually the case, the project to combine these publications proved more difficult than originally anticipated. However, the result was truly a hybrid of the earlier works, and yet also a new publication.

The University of Vermont Extension System printed 3,000 copies of the first edition of this work in 1995. In less than three years, the entire inventory was depleted. Much to our surprise, the work was used widely outside of Vermont. This second edition, published by the Natural Resource, Agriculture, and Engineering Service (NRAES), has been prepared for a broader audience. Virtually all of the concepts apply not only to the northeastern states, but to the entire area from West Virginia to Maine and from Massachusetts to Minnesota. And, although the forest communities, species, and site descriptions may look foreign to those outside the northeastern states, the concepts of ecology and silviculture still apply.

Chapter 1
Introduction to Forest Ecology

Forest ecology is the study of life in areas where the predominant vegetation is trees. The word ecology is derived from the ancient Greek word *oikos*, meaning "house," and the suffix *ology*, which means "study of." The central idea of viewing a community of plants and animals — that exist in the same place and time — as a "house" is crucial to understanding how ecology relates to forest management practices and timber harvesting. Just as members of a household influence one another, changes in the forest influence the plants and animals that live there. Human use of forests can have major impacts. For this reason, it is important that woodland owners, loggers, and foresters understand the nature of changes brought about by management practices — especially timber harvesting — and how to work with forest ecosystems, rather than against them.

As a discipline, forest ecology is very broad and theoretical. From understanding how one population of soil bacteria relates to another, to characterizing the potential of different sites for growing **northern hardwoods** (words appearing in bold italics when first mentioned are defined in the glossary, beginning on page 73) — and everything in between — forest ecology deals with virtually anything that relates to the plants and animals in woodlands. And, because so many different organisms in forests influence one another in complex ways that vary depending on local site conditions, the application of forest ecologists' work is often based more on theories and trends than on facts supported by rigorous scientific analysis.

The study of forest ecology is one of the most important courses taught in forestry schools. It forms the basis for later courses, especially silviculture — the art and science of controlling the species mix, growth rate, size, and *form* of trees in forests for the production of wood products and other benefits.

The content of this publication is designed around three main ideas. First, *practical* concepts are emphasized, but not to the point of oversimplifying the subjects to avoid a theoretical discussion. Second, although the subjects of ecology and silviculture are filled with jargon, I have avoided using specialized terms as much as possible. Where jargon cannot be avoided, definitions for terms are located in the glossary, beginning on page 73. Finally, the material is presented in a way that is relevant to as many differ-ent readers as possible, but especially to woodland owners, log-gers, and foresters. Subject depth is the victim of breadth, since forest ecology and silviculture are huge subjects to cover in such a small book.

The goal here is to convince readers that timber harvesting is a *disturbance* that can be controlled, to some degree, with knowl-edge about how the forest will react to changes. Often the eco-nomic constraints of timber extraction result in harvesting prac-tices that follow the path of least resistance. Even so, an indi-vidual who understands forest ecosystems will be a better log-ger, forest owner, or manager.

The Forests of the Northeast

More than half the land area of the Northeast is covered with forests, almost all of which are capable of growing timber. Three major *forest regions* in North America converge in the North-east: from the extreme north, the boreal forest region; from the Northeast's middle latitudes, the northern hardwood forest; and from the Northeast's southernmost latitudes, the central hard-wood forest (figure 1.1). These forest regions form the basis of more than thirty *forest types*, or associations of tree species, rec-ognized by the Society of American Foresters. Twenty-one of these types include timber species that are important to our economy (see appendix A, page 70).

By far, the most extensive forest regions are the central and north-ern hardwood forests, and the most extensive forest types are associations of hardwoods. In the north, the most extensive type is sugar maple–beech–yellow birch, and the most important tim-ber species are sugar maple, yellow birch, and white ash. In the central and southern portions of the area, the central hardwood forests predominate. The most extensive type is oak–hickory, and

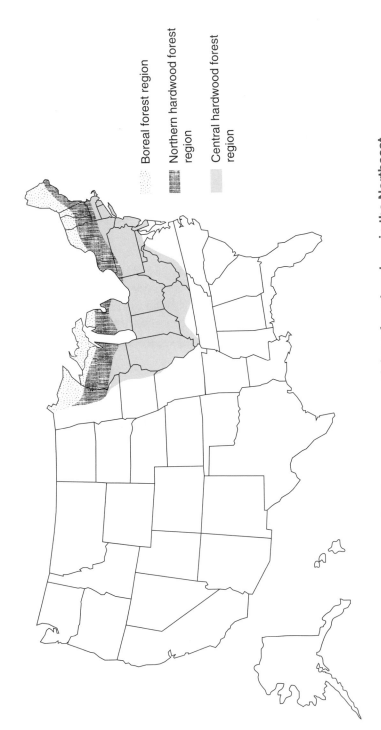

Figure 1.1 A generalized map showing the convergence of three forest regions in the Northeast

Boreal forest region

Northern hardwood forest region

Central hardwood forest region

the most important species represent members of the red oak and white oak groups. Where the two regions meet, there are many variations of the different species that represent the predominant types. Generally, as one proceeds north, forest complexity and species diversity decrease. An exception to this generalization is that where forest types mix, species diversity is high.

Forest regions and types are influenced by a number of different factors, but climate — especially rainfall and temperature extremes — has a decided influence on what grows where. Within a region, however, forest types are determined by a complex interaction of factors, including soils, *slope* and *aspect,* elevation, and local climatic conditions.

Figure 1.2 illustrates the influence of topography on local climate. As one ascends, he or she can expect to encounter forest types that are more typical in northern latitudes. For instance, mountain-top forests in the southern portion of the region at 4,000–5,000 feet elevation are often composed of boreal forest species.

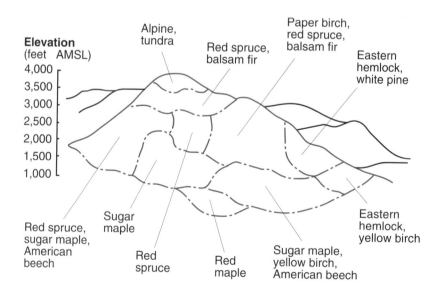

Figure 1.2 The influence of topography on local climate

Think of a forest type as a local expression of a forest region, where the effects of climate are combined with other, site-related factors. The forest *stand* is the smallest division recognized in forest management. According to Dr. David Smith in *The Practice of Silviculture*, a stand "is an area of forest where the trees are sufficiently uniform in age, structure, or species composition as to be distinguishable from surrounding areas." (For more information on this and other works referred to, see "References," beginning on page 85.)

As classification of forest vegetation proceeds from forest regions to forest types to stands, the complexity of factors that separate one type or stand from another increases. Even within stands, variation can be tremendous. Think of forest stands as local forest types that have been disturbed. Depending on the timing, severity, and type of disturbance, the species composition of the existing stand may only be temporary and the potential vegetation quite different from the trees that will predominate, with or without subsequent disturbances. For instance, a recently disturbed northern hardwood forest may be mostly composed of aspen and white birch. Given time and respite from severe disturbances, the future forest will evolve into an association of species that is completely different. The species that comprise a given forest type are usually temporary, and disturbances will change the picture. This is a fairly simplistic description, but the main point is that forest stands change with time and are the result of disturbances of a forest type. Forest types are a localized expression of a forest region, which is primarily determined by climatic conditions. Therefore, knowing the history of disturbances is necessary in order to understand existing forests and how they will change over time.

Chapter 2
Forest Site

The sum of all the natural factors that influence forest growth is characterized as the "site." Foresters often talk about *site index*, or good sites versus bad sites, or hardwood sites versus pine sites. What they are referring to is the combination of environmental factors such as climate, soils, slope and aspect, and elevation, all of which affect not only which trees grow where, but how fast they grow and what their form at *maturity* will be. Tree form refers to the degree of taper between the stump and the top. Taper is usually more extreme on poorer sites.

Slope, aspect, and elevation are factors which, in combination with local climatic conditions, determine the microclimate, or localized weather conditions, of a site. For example, northeast-facing slopes tend to be cool, moist, and more productive than southwest-facing slopes, which tend to be warm and dry. Higher-elevation sites are cooler and more exposed than lowland sites. Average temperatures are about 5 degrees cooler per 1,000 feet elevation. Just by knowing the aspect, elevation, and slope of a site, one can get a good idea about its potential *productivity* and which species will grow well there.

The concept of site is one of the most important in forestry because the interaction between a stand and its site determines growth potential. Although a combination of factors always determines site quality, all other factors being equal, site quality for timber production is largely a function of soil quality, especially the physical, chemical, and biological properties of soil.

Soils

The characteristics of a forest soil are defined by varying combinations of four main ingredients — mineral particles, organic matter, water, and air.

Chemical properties are determined mostly by the type of minerals from which the soil is derived. For example, limestone-derived soils tend to be more fertile, while soils derived from granite tend to be infertile. Fertility is largely a function of a soil's chemical properties. *Soil pH*, a measure of the acid-intensity of a soil solution, is a common measure related to fertility. Forest soils tend to be acidic, but, in general, the more acid a soil, the less fertile it is.

The physical properties of a soil — whether it is fine or coarse — depend not only on the type of minerals present, but also the particle size. Fine soils, with a higher content of silt and clay, tend to be more productive than coarse soils, with a higher content of sand and gravel. Fine soils, however, usually do not drain as well and are more easily eroded. Fine soils are also more susceptible to damage from compaction, especially when wet.

The presence of organic matter — decaying debris such as leaves and twigs — influences both the physical and chemical properties of a soil. Soils with a high proportion of organic matter tend to have better structure (more aggregation of particles) and are more fertile. Organic matter also improves the soil's chemistry and water-holding capacity.

Over 50% of the volume in the upper layers of an undisturbed forest soil is made up of pore space that holds air and water. The proportion of air and water is both affected and determined by the physical properties of the soil. Since roots need air to breath and water to supply the rest of the tree, these physical properties of soil are extremely important. Activities that compact soil will eliminate space for air and water and lower the productivity of a site. This is why it is so important for logging equipment to stay on established trails, especially on sites with finer-textured soils that tend to compact more readily. Consider, too, that more than half of the *feeder roots* in a forest are found in the top six inches of soil.

The biological properties of a soil include the communities of fungi, bacteria, and insects that live on organic matter produced by trees and shrubs. Though mostly microscopic, these organisms are absolutely essential to the growth and development of forests. Occasionally some can cause disease in trees, but most feed on fallen leaves and woody debris. The main role of such organisms is to recycle nutrients, such as phosphorus, potassium, and calcium, that are tied up in vegetation and other organisms. Without these decomposers, the forest floor would be littered with debris accumulated over thousands of years, and trees would lack the essential elements locked up in this material.

Some organisms, such as a group known as mycorrhizae-forming fungi, cause beneficial infections on tree roots. In exchange for a share of the energy manufactured by the tree, these fungi tremendously improve a tree's ability to take up water and nutrients. The tree and mycorrhizae-forming fungi depend on one another, each supplying the other with sustenance for a small cost. Unfortunately, mycorrhizae roots are very susceptible to changes in environmental conditions, especially those caused by soil compaction. For this reason, forest management decisions must take into consideration the way soil organisms will react to logging. Usually their activity increases due to higher temperatures on the forest floor when a portion of the *overstory* is removed. Soil compaction, however, causes drastic changes in soil chemistry and will inhibit mycorrhizal growth.

Nutrient Cycling

Of the twenty elements that are required for plant growth, five are taken up from the soil in relatively large quantities — nitrogen, phosphorus, potassium, magnesium, and calcium. According to recent research, on most forest sites all five elements are imported through rainfall in quantities sufficient to sustain *sawtimber* harvesting when the *rotation* exceeds thirty years (Wegner, 1984). Since the imported quantities of these elements are relatively small (only a few ounces of phosphorus per acre per year, for example), nutrients must be recycled through the forest floor. Soil organisms are responsible for conserving nutrients that are later taken up by trees and other vegetation.

When harvesting includes more of the tops of trees, the quantity of nutrients exported from the site increases dramatically. The finer the material removed (branches, twigs, and leaves) in addition to the *stem*, the more nutrients taken from the site. This reflects the fact that, in the aboveground parts of a tree, the highest concentrations of nutrients are in leaves and twigs. Notwith-

Introduction to Forest Ecology and Silviculture

standing, whole-tree harvesting — except on the poorest sites — should not substantially reduce site productivity for wood fiber during a single rotation. Over the course of two or three rotations, export of some nutrients may exceed the capacity of the site to replenish them (Pritchett, 1987).

Characterizing Site Quality

Assessment of site quality helps to predict timber *yield* and to set forest management priorities. Where timber production is a primary objective, sites that have the greatest potential to respond to *treatment* are usually scheduled first. Treatment costs are recovered sooner in a stand on a good site that will respond rapidly to a *prescription* (also see "The Purpose of Silviculture," pages 37–43).

Trees grow in temperate regions by putting down a new layer of wood each year. On the main stem, this new wood increases the girth of the tree, but in the top of the tree the same layer elongates the stem, increasing total tree height. Diameter growth rate is controlled by stem density; the more crowded it is, the slower trees grow in diameter. When stands are crowded, a higher proportion of wood is devoted to elongating the top of the tree, probably to improve its chances of putting leaves in sunlight. When stands are not crowded, trees will lay down proportionally more wood lower on the stem, possibly as a way of improving transport of water and nutrients to leaves. This relationship between top elongation and girth means that if tree density is held constant, the height of a tree expressed as a function of its age is a good indicator of site quality.

Within a species, the relationship of height to age is relatively constant throughout the life of a stand. On a "good" site, for example, one can expect to find trees 30% taller than trees of the same age on a "poor" site, and the relationship holds true from *sapling* to sawtimber. However, since it is unlikely a forester will encounter trees of the same age on different sites to allow a direct comparison, in the East we use an index based on the height, or predicted height, of trees when they are fifty years old. Known among foresters as site index, it is based on the direct measurement of total tree height and age (obtained from a count of annual rings at breast height), but expressed as the predicted height at an index age of fifty years. For example, a tree that is 60 feet tall at the index age of fifty years is said to have a site index of 60. That same tree at thirty years old has the same site index, and one would expect it to be 60 feet tall when the tree is fifty years old. Site index curves allow the forester to use total height and age to predict site index for almost any combination of tree heights and stand ages.

One limitation of site index is that it is species-specific, so it is correctly stated as "sugar maple site index," or "red oak site index." There is sometimes a basis for comparison across species, but not always. An excellent site for red oak may prove only an average-to-good site for sugar maple, and a great pine site is usually not as good a site for most hardwood species. A good soil is a good soil, but site index integrates a number of other factors that influence production. Site index is more comparable among species that are common associates — sugar maple and yellow birch, for instance — than between species that do not usually grow together.

Another limitation of site index is that the user assumes trees have not been suppressed, but that they have always had competition from other surrounding trees (which is the same as saying the stand has always been fully stocked). The concept of site index does not apply in *uneven-aged* stands or in stands that have had drastic changes in stocking due to earlier cutting, or insect- and disease-related mortality. The most common error is overestimating productivity, predicting a site index that is higher than the site deserves. But since site index is mostly used on a relative scale, to say one site is better than another for a particular species, precise measurements are rarely necessary.

Site index evolved as a way for managers to quickly identify productive forest sites around the turn of the century, when most of the woodlands in the East were young sapling stands recovering from heavy cutting to fuel foundries and use for building. In the 1930s, Luther Schnur, a silviculturist for the U.S. Forest Service, developed a way to use site index to predict *yield* in upland oak stands. Yield is the production of usable wood fiber per unit time; the higher a site index, the greater the yield. Tying site index to yield was no easy matter, especially since he had a great deal of data for young stands, but almost no information on older stands. The result was that although Schnur proposed a very useful tool to assess relative forest productivity, his lack of data points in older stands made it almost impossible to predict yield later in the life of a stand. Today we use site index to determine the range of potential productivity in a particular forest and to help prioritize *silvicultural* decisions.

Site index is typically measured as follows: total tree height is estimated using a clinometer and principles of basic geometry (similar triangles). A tool called an increment borer (figure 2.1) is used to remove a straw-sized piece of wood at *"diameter breast height"* or *"diameter at breast height"* (*d.b.h.*) that includes growth rings from the current year back to when the tree was 4.5 feet tall. Almost all measures in forestry are based on d.b.h. because it eliminates the effect of flaring where the stem meets the ground (also known as butt swell). Extracting an increment core damages the wood, but if done correctly, it does not injure the tree. Unfortunately, it is the only way of aging hardwoods, short of cutting the tree down. White and red pine are easy to age without boring since they produce one *whorl* of branches each year (a whorl is a group of branches that come out from the

Figure 2.1 An increment borer is used to take a sample in order to age a tree.

stem all at the same height, separated by clear stem for a foot or two until the next year's whorl). It is a simple matter of counting branch whorls to the top of the tree. The estimates for total height and age are plotted on an appropriate set of site index curves to obtain a value for site index.

Site-index curves are species-specific and based on the height of free-grown forest trees at an index age of fifty years. Figure 2.2 shows site-index curves for sugar maple in Vermont and New Hampshire. A sugar maple that is thirty-five years old and 55 feet tall is on a site with a sugar maple site index of 70; when the tree is fifty years old, it should be about 70 feet tall.

Tree form (taper of the stem), bark appearance, and species composition of the stand are three good indicators of site quality. Although a straight tree — with a slightly tapering stem and a tall, full crown — may be an inherent genetic characteristic of a

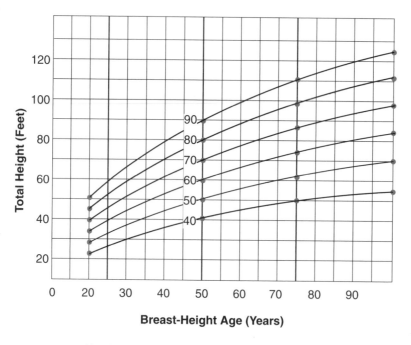

Numbers above curves refer to site-index values.

Figure 2.2 Sample site-index curves for sugar maple in Vermont and New Hampshire

Introduction to Forest Ecology and Silviculture

species, it is more apt to be a full expression of the capabilities of the tree on a favorable site. The very same tree on a poor site will look totally different, old before its time and dwarfed.

For some species, bark characteristics are an excellent indicator of site quality. Sugar maple, white ash, and red oak, in particular, but other species to a lesser degree, have bark that develops telltale signs of rapid growth: when young sawtimber on a good site begins to grow rapidly, it looks as though the bark is being stretched.

An understanding of the nature of forest sites is absolutely essential to making sound forest management decisions, but it is only part of the picture. Site productivity is influenced by management activities. However, short of practices such as fertilization that because of their expense are largely impractical today, forest sites cannot be improved. On the other hand, sites are easily damaged by careless logging, excessive foot traffic, and poorly designed roads and *skid trails*. For these reasons, the purposes of timber management are to 1) minimize the damage caused by periodic harvesting; 2) adjust species composition to suit the site; and 3) allocate site resources, by *thinning*, to selected trees that have the greatest biological and economic potential.

Sites are also characterized by their potential to provide habitat for different species of wildlife. Although not as easily quantified as site index, habitat opportunities are easy for the experienced eye to see. When providing for wildlife is a primary objective, it is usually easy to adapt timber-oriented practices to suit the circumstances. Only rarely does an emphasis on one goal or objective exclude another. Managing for timber and improving or creating *wildlife habitats* are very compatible objectives in northeastern forests.

Chapter 3
How Forests Grow and Change

Basic Forest Tree Biology

Trees and other plants are often referred to as factories because they convert light energy into chemical energy. Through a process called photosynthesis, the leaves capture solar energy by converting carbon, hydrogen, and oxygen into complex sugars — the products of photosynthesis. One of these complex sugars is cellulose, the main ingredient of wood fiber.

Although a complete knowledge of photosynthesis is not essential to understanding how trees grow, it is important to realize that, in theory, the faster and more efficiently a tree carries on photosynthesis, the faster it will grow. Availability of light and of water are two factors that are controlled by harvesting. When a portion of the trees in a stand is removed, the photosynthetic potential is reallocated to the remaining trees. In a young stand, it is usually only a matter of five to ten years before the crowns of *residual* trees grow into the spaces left by those taken out.

Trees are composed principally of four main parts: roots, stem, branches, and leaves. Other specialized structures, like flowers and seeds, develop periodically for purposes of reproduction. However, virtually all of the important physiological processes in trees involve one or more of the four main parts.

About 20% of the mass of a forest-grown tree is devoted to roots. In addition to anchoring the tree, roots gather mineral nutrients, take up water, and store the products of photosynthesis. Forest tree roots are much more extensive than they appear. For example, the root system of a sugar maple may extend as much as two to five times beyond the spread of a tree's *crown.* Most of these roots, known as fine or feeder roots, are within a few inches of the surface; and, though the fine roots may account for only 14% of the total root mass, 80% of total root length is in fine roots. Consequently, roots are everywhere in the forest, and the ones that are most sensitive to damage are also the most susceptible to damage and the most crucial to tree health.

The main stem usually makes up about 60% of a tree's weight. The stem supports the leaves and branches and serves as the main plumbing system, with vessels to transport water and nutrients up to the leaves and other cells that transport the products of photosynthesis to living tissue throughout the tree. The growing portion of a tree is only a thin layer of cells surrounding the main stem. Each year this thin sheath of cells puts down a new layer of *sapwood* and bark. The rate at which cells are laid down determines how fast a tree grows in girth.

A tree's branches support leaves in a configuration that maximizes light availability or protects the leaves from excessive exposure on harsh sites. Branches also serve as the second-order plumbing of the main stem. The leaves carry on photosynthesis and exchange important gases, such as oxygen and carbon dioxide, with the atmosphere. Combined, branches and leaves make up about 20% of the tree's total weight.

All trees have roots, stems, branches, and leaves, but the form of each of these components differs among species — and within a species in some cases. Some characteristics, such as the size and shape of leaves on the top of the crown versus those on shaded lower branches, even differ on the same tree. For this reason, leaf size and shape are said to be plastic because they are characteristics that mold themselves to the circumstances. This is one of the reasons why learning to identify trees solely by their leaves is so unreliable and often frustrating.

Although many of the obvious differences among individual trees within a species are random, some are not. Important differences in a tree's form and function may be caused by environmental conditions. For example, the root systems of most trees tend to be more extensive on drier sites. Another example is that *open-grown* trees tend to have short trunks and wide, deep crowns, while forest-grown trees of the same species, in their struggle to obtain crown space, tend to have long boles and short irregular crowns that fit the available space in the *canopy.*

Important genetic differences between species have evolved over millions of years. Not all structural differences are due to adaptations that make one species a better competitor than another on a given site, but many structural differences are. For example, many conifers have adapted to become better competitors on dry, exposed sites than most hardwood species.

Though most conifers will do well on "hardwood" sites, their natural habitat is defined by the limits of *tolerance* of other species. White pine on a hardwood site is a good example. It grows extremely well on moist, protected sites, but it is nearly impossible to get a new stand of pine started using natural *regeneration* following logging. Hardwood species, such as sugar maple and yellow birch, are much better competitors in the *understory*. On a drier site, the reverse may be true — white pine can compete more effectively than most hardwoods (table 3.1).

Each species has a range of environments in which it will grow. These extend from circumstances where a given species is a minor component, poorly formed and slow growing, to situations where a tree is able to take full advantage of a site and grow to its maximum biological potential. One of the most common silvicultural errors in forest management is trying to grow a species on a site where it can achieve only a fraction of its growth potential.

Forest Succession and Tolerance

In the absence of disturbance, changes in the species composition of a forest are slow, but continuous. The process of continual change is called *succession*. The fact that the direction of forest succession is both predictable and controllable is the foundation of forestry practice. Succession can be speeded up, slowed down, or stopped altogether simply by altering the composition and density of trees in a stand. Fire, hurricane, disease, or *high-grading* can have the same effect, but forest succession is purely chance without careful planning.

Primary succession is the progression of plant and animal communities from the colonizers of bare mineral soil to a relatively stable, self-sustaining community many years later. In a very broad context, forests in the glaciated portions of North America

Table 3.1 Minimum site requirements for adequate growth of selected forest trees

Species	Moisture	Nutrients	Heat	Light
Balsam Fir	H	M	L	M
White Spruce	M	M	L	M
Red Pine	L	L	M	V
White Pine	L	M	M	M
White Cedar	V	M	L	L
Larch	V	L	L	V
Red Cedar	L	M	V	H
Red Maple	L	M	M	H
Sugar Maple	H	V	H	L
Silver Maple	H	V	V	V
Yellow Birch	H	V	M	L
Paper Birch	M	M	L	V
Butternut Hickory	M	V	V	L
White Ash	H	V	V	H
Black Walnut	L	V	V	M
Ironwood	M	V	V	L
Cottonwood	M	V	V	H
Aspen	M	M	M	V
White Oak	M	V	L	L
Red Oak	L	V	H	M
Basswood	M	V	H	L
Elm	H	V	H	M

Key: L = Low, M = Medium, H = High, V = Very high

Source: Adapted from Beattie, Thompson, and Levine

are the result of primary succession on pure mineral sands and cobbles left by glaciers when they last retreated more than 10,000 years ago. *Secondary* succession takes place after disturbance in an existing plant community. It is an interruption in a primary successional cycle. Forest management, then, deals principally with secondary succession. Figure 3.1. (page 18) shows a secondary successional pattern for northern hardwoods in northern New England. This pattern will vary due to the nature of disturbance, soils, exposure, elevation, and other site-related factors.

Of the eight or ten stages of primary succession recognized by ecologists, only five apply to forest management. They are herbaceous, shrubs, *intolerant* trees, midtolerant trees, and toler-

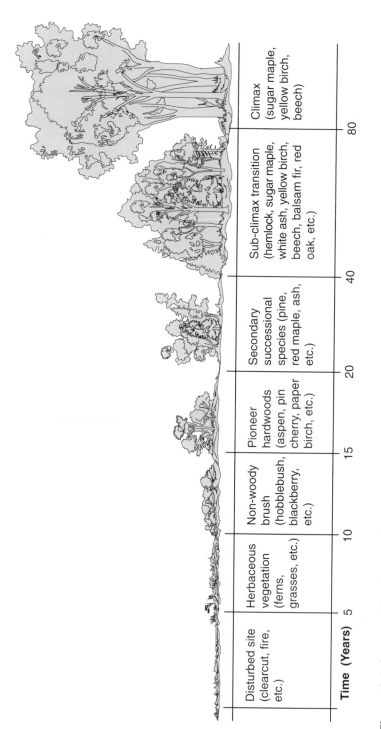

Disturbed site (clearcut, fire, etc.)	Herbaceous vegetation (ferns, grasses, etc.)	Non-woody brush (hobblebush, blackberry, etc.)	Pioneer hardwoods (aspen, pin cherry, paper birch, etc.)	Secondary successional species (pine, red maple, ash, etc.)	Sub-climax transition (hemlock, sugar maple, white ash, yellow birch, beech, balsam fir, red oak, etc.)	Climax (sugar maple, yellow birch, beech)

Time (Years) 5   10   15   20   40   80

Figure 3.1 A generalized secondary successional pattern for northern hardwoods in northern New England

ant trees. Intolerant trees are also known as *pioneers*; midtolerant trees as *early-successional* or "sub-climax"; and tolerant trees as *late–successional* or "climax" species. As a site proceeds from pioneer to climax, the complexity of the ecosystem usually increases and its stability or resistance to change — in the absence of disturbance — increases as well.

Pioneer species, such as aspen and white birch, grow fast but are short-lived. Their size at maturity is usually much smaller than climax species such as sugar maple. However, the most distinctive difference between pioneer and climax species is that pioneer trees are incapable of establishing themselves in a forest understory; they are therefore said to be intolerant.

Though tolerance is really the degree to which a species can share resources on a site and still be successful, it is most often thought of as shade tolerance. Table 3.2 compares tolerance to low-light conditions for selected commercially important trees. Pioneer species have virtually no tolerance of shaded conditions, while climax species do. As a result, forest succession in the absence of disturbance proceeds toward more tolerant species. Since toler-

Table 3.2 Relative tolerance of fifteen commercially important trees

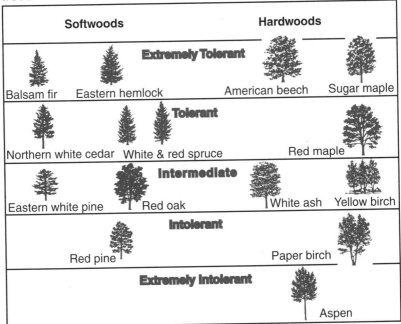

ant species can establish themselves in shade, they remain in a forest stand in the absence of disturbance, creating a condition known as a climax forest.

An understanding of species tolerance is fundamental to forest management. Pioneer stands that are thinned will move toward a climax association more quickly than if left alone. By the same token, climax stands that are lightly thinned will continue to be climax stands. Pioneer stands that are clear-cut will come back as pioneer stands, and climax stands that are clear-cut usually result in early-successional stands. **The more drastic the disturbance, the further back succession is set.** For example, if the prescription calls for regenerating pioneer or early-successional species such as aspen or white birch, a logger might be encouraged to disturb the site as much as possible. The small, usually wind-borne seeds of pioneer species require a bare, mineral-soil seedbed to germinate well. Also, the root systems of some pioneer species such as aspen will sprout new stems called *suckers* when the stand is clear-cut. This is how aspen got its reputation as "the tree that loves to be hated"!

In some instances, where a given species falls in forest succession depends upon the site. White pine is a good example. Usually intermediate in tolerance, on good hardwood sites it acts like a pioneer species, but on drier "pine" sites it acts more like a climax species. So tolerance, and the speed and direction of succession, are often tied to site characteristics. Change in forests is the rule rather than the exception. Predicting the speed and direction of change is what forest management is all about.

Silvics of Some Important Timber Species

Over millions of years, trees have evolved characteristics that enable them to reproduce, establish, grow, and maintain themselves within a range of environmental conditions. Some species, such as red maple, are tolerant of a wide range of conditions, while others, such as white ash, have a much more narrow range. Forest management attempts to complement a species' characteristics, or its strategy, with the site in question. The driving forces behind timber management prescriptions are knowledge of the *silvics* — or life histories and characteristics

— of important timber species and experience with local variations and how they interact with different sites.

There are more than twenty commercially important timber species in the Northeast, too many to cover here. However, four — aspen, eastern white pine, northern red oak, and sugar maple — will illustrate succession, tolerance, and the range of silvical characteristics among the major timber species. More detailed silvical descriptions of these and other North American species can be found in Fowells, ed., *Silvics of Forest Trees of the United States.*

Aspen

Aspen is the most widely distributed tree species in North America. (There are two species — quaking and bigtooth — that are so similar in the Northeast they are both referred to as aspen.) Also known as popple, aspen is a very intolerant pioneer species, which will take hold in almost any bare soil, but does best on fertile, well-drained sites.

Although aspen, like many pioneer species, produces heavy crops of light, wind-blown seeds — usually every four to five years — it reproduces most vigorously from "suckering" off of the roots of trees that have been cut. As a result, whole stands of aspen are often composed of multiple stems of a few individuals, also known as *clones.*

Aspen is very fast growing, but short lived (table 3.3, page 22). In the Northeast, clones much over sixty years old are rare. Usually no larger than 10 to 14 inches in diameter and 50 to 60 feet tall at maturity, trees older than sixty years deteriorate rapidly from disease and insects. Aspen is also one of the most important food species for wildlife in the region, especially for the ruffed grouse, which will feed on its buds (the large flower buds of male trees are a favorite) throughout the winter. Deer and beaver also rely on aspen for food.

The only way to maintain aspen in stands where forest succession is moving beyond the early stages is to clear-cut the entire clone and surrounding vegetation. Aspen requires an opening large enough for sun to warm the ground (the diameter of openings should be at least one and one-half times the height of sur-

Table 3.3 Total anticipated height of quaking aspen by age on sites of varying productivity in the Lake States

Age (Years)	Total Height in Feet on the Following Sites*				
	Excellent	Good	Medium	Poor	Off-Site
20	44	39	34	28	23
30	59	51	44	37	29
40	70	62	53	44	35
50	79	69	60	50	39
60	87	76	66	55	
70	93	82	70	59	
80	98	86	74		

*Average height for dominant trees in well-stocked, even–aged stands

Source: Fowells, ed.

rounding trees). Sprouting vigor decreases with age, and shaded conditions following harvest will lessen sprouting. Best success follows dormant-season, frozen-ground logging conditions. Unless a prescription calls for removing an aspen overstory to favor an understory of more tolerant species, aspen stands should not be thinned. Doing so will speed up forest succession and eliminate aspen from the stand. In Minnesota, however, pure young aspen stands are thinned to improve diameter growth rates of residual trees.

Eastern White Pine

White pine is most commonly associated with coarse, well-drained soils, but it will also grow extremely well on more fertile sites. Usually competition from understory hardwoods on good sites will successfully block establishment of white pine regeneration.

Though considered a sub-climax or even climax species on some sites — especially on coarse-textured soils — white pine has only "intermediate" tolerance. It can outcompete aspen and birches, but has a difficult time with late-successional northern hardwoods such as American beech, sugar maple, or yellow birch. Much of the pine in the Northeast today got its start on abandoned pasture where, in many instances, it was a pioneer.

Like most conifers, white pine relies on seed for reproduction. Cones take two years to mature and good seed years occur every five to ten years. Planning regeneration prescriptions around good seed years is essential with white pine. Seeds that find bare mineral soil germinate best. Once established, eastern white pine is a long-lived tree that can attain immense proportions on good sites (table 3.4).

One of the principal enemies of white pine is the white-pine weevil. This tip-feeding insect kills the terminal shoot and causes stem deformation as well as growth reduction. Research suggests that white pine allowed to grow to *pole-timber* size with an overstory present is less susceptible to weevil damage.

Northern Red Oak

Northern red oak grows on a variety of sites and does extremely well on fertile soils. In northern New England, red oak is mostly found on warmer sites, often those with south- and west-facing aspects.

Red oak is intermediate in tolerance. It will establish in partial shade, but will not persist unless the overstory opens up after

Table 3.4 Estimated yields of fully stocked white pine stands on sites of varying productivity

Age	Site Index (MBF)			
(Years)	45	55	65	75
50	8	18	28	38
100	40	52	68	78

Estimated Yields of Fully Stocked Pure White Pine Stands

Source: Fowells, ed.

establishment. It does not usually produce seed until thirty to forty years of age. After that, good seed years come at two- to three-year intervals. Seeds germinate best in mineral soil with a light layer of leaf litter. Acorns, however, are a popular food source for many animals and the bulk of most seed crops goes to wildlife and insects. Some animals, such as blue jays and squirrels, are believed essential in seed distribution because of their habits of gathering and storing acorns. Red oak also regenerates by *stump sprouts*. As with aspen, sprouting vigor declines with age.

Red oak grows fast on a good site. Maximum diameter growth begins at about the time trees reach small sawtimber size. Harvesting oak at this time, except for the purposes of light thinning (to which the stand will respond very well) is not a good practice. Increase in log values through this period of accelerated growth can be five to ten times the rate of volume increase. Though stands may reach *merchantable* size by age sixty, the optimum rotation is usually eighty to a hundred years or more. Generally, the better the site for red oak, the longer the rotation.

Sugar Maple

Sugar maple is one of the most important tree species in the region, especially in New England. It grows in pure stands, and is also a major or minor part of at least a dozen other forest cover types. As a climax species, sugar maple is extremely *shade tolerant* and will respond to thinning after many years in the understory. It will grow just about anywhere, except at high elevations and on wet sites, but thrives on moist, well-drained, fertile soils below 2,500 feet in elevation.

Sugar maple produces seed each year, but good crops come at two- to five-year intervals. Though it will sprout from the stump, like aspen and red oak its sprouting vigor decreases rapidly with age. With sugar maple, reproduction from seed is usually not a problem, although establishment of a new stand in the understory is more difficult. Very light, frequent cutting will favor sugar maple more than opening a stand with a closed canopy all in one operation.

On an average site, a 35-year-old tree is about 40 feet tall and 6 to 8 inches in stem diameter. Most stands reach maturity at 100 to 150 years, after which height and diameter growth are negligible. Even though sugar maple trees can inhibit the spread of infection following wounds from logging and tapping, the amount of defect in unmanaged stands can be as high as 35–50%. In a study of lumber yields from trees that had been tapped for maple sap production, however, Sendak (1982) discovered that the average value-loss per tree was only $2.87.

Chapter 4
The Effects of Stress and Disturbance on Forest Ecosystems

A forest is more than trees. It is a complex ecosystem — ever changing and defined by the interactions of living organisms and the surrounding environment. For this reason, management decisions should consider the potential impacts on the whole forest. Managers have come to realize that the forest is more than just the sum of its parts.

An ecosystem can be characterized at any scale, from a few square feet to thousands of acres or the entire earth. Wherever plants and animals interact with their environment and each other and the ultimate source of energy is sunlight, the association can be described as an ecosystem. For example, the interaction of lichen populations on the bark of a beech tree is as much an ecosystem as the community in which the beech tree is found. It is just a different scale. (Also see chapter 12, "Ecosystem Management," beginning on page 67.)

The study of ecosystems is mostly concerned with the way different species interact, changes in the community over time, and the flow of inputs and outputs, such as energy, nutrients, and water. Even the simplest ecosystem can be tremendously complex. Change any part, and it influences the rest of the system.

In science class, young people learn about the interdependence of different parts of an ecosystem through an exercise called the web of nature. Students are arranged in a circle and each is assigned to play a part of the ecosystem — sunlight, nutrients, water, and other components. The instructor begins connecting the students with a ball of string. After all possible connections are made, there is a web among them. Then, to illustrate the effects of disturbance in the ecosystem, the teacher begins removing students from the circle. The effect, of course, is a slackening of the web. The students reposition themselves to take up the slack, and the result is usually something other than a circle. As

shown in figure 4.1, when an ecosystem, such as a greatly simplified five-factor ecosystem (A), is disrupted by stress or disturbance, the way important factors interact is changed, effectively throwing things off balance (B). Even though the unbal-
ancing may lead to permanent changes in the ecosystem — which
may be good or bad depending on one's perspective — eventually the system will reach a new equilibrium (C).

Forests react to changes in the same way. Almost regardless of
the change, whether caused by harvesting, hurricane, disease,
or insects, the forest ecosystem eventually takes up the slack by
establishing a new equilibrium. Depending upon the severity of
disturbance, a new equilibrium may cause changes in soil organisms, wildlife populations, or productivity and composition
of tree species. The direction and degree of change is a result of
the way organisms react to one another and to the new condi-

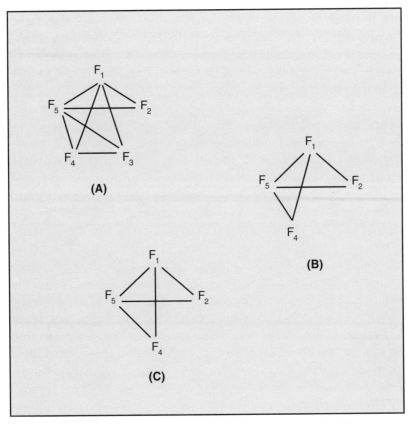

**Figure 4.1 Diagram of interdependence in a simplified five-factor
ecosystem**

tions on the site. The new balance can be desirable or, as in the case of human-caused deforestation in tropical forests, potentially disastrous.

Stress is caused when some important ecological factor changes, resulting in strain to reach a new equilibrium. Like a motor that burns oil, as oil pressure drops, the engine runs hotter, straining the cooling system. In forests, stress is the rule rather than the exception, and not all stresses are bad. For instance, dry conditions can lead to stress in trees caused by a lack of moisture. Under these conditions, leaves will wilt during the hottest part of the day. Wilting helps leaves conserve water and it is a normal reaction to moisture deficit that is not damaging to the tree. But the moment stresses are compounded (like a hotter engine burning more oil), growth reduction, crown *dieback,* and *mortality* are apt to occur. As figure 4.2 illustrates, disturbance and stress are related phenomena; generally, the effect of one is to cause the other. Single disturbances or stressors are not necessarily bad, but when compounded with other stresses (S_2 and S_3) or further disturbance (D_2 and D_3) the compounding of effects can lead to crown dieback and mortality. Dead branches in the

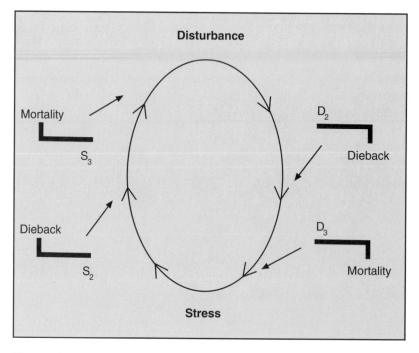

Figure 4.2 Disturbance and stress are related phenomena.

Introduction to Forest Ecology and Silviculture

tops of crowns are danger signs; the death of lower branches is usually due to natural causes.

Climate-induced disturbances, such as drought, hurricanes, and ice storms, are expected but unpredictable events. For example, every hundred years or so New England forests will suffer extensive damage from hurricane and other wind events. Other natural disturbances include those caused by disease and insects. The presence of some of the most serious pests, however, is due to human error. Chestnut blight, Dutch elm disease, and gypsy moth are three examples of seriously destructive pests that were imported by humans. Each has had a tremendous impact on the species composition of forests in the Northeast. For example, we would have only a fraction of the oak that we see in the region today if it were not for the chestnut blight.

In one of the most devastating winter storms anyone in the Northeast can remember, nearly 25 million acres of forests — from northwestern New York and southern Quebec to the south-central Maine coast — were coated with ice over the course of three days in early January 1998. It was a very unusual meteorological event in terms both of the way the storm evolved and the extent of impact. Tree damage throughout the region consisted of broken limbs and crowns, and in some areas the ice was so heavy that trees were uprooted or main stems snapped. In the Champlain Valley of Vermont, which was hit hard, ice loading on branches caused twig weights to increase 16 to 159 times their normal weight, according to Dr. Dale Bergdahl, a forest pathologist at the University of Vermont. This one storm has had a severe and long-lasting effect on forests of the region. Although it is not apt to change the species composition of the affected forests, these stands are now much more susceptible to secondary infections from disease-causing fungi. And product values in some especially hard-hit areas have plummeted from the effects of ice.

Pear thrips is an emerging insect that could have serious consequences for sugar maple in the Northeast. Virtually unknown as a pest in forests as little as fifteen years ago, it defoliated more than 400,000 acres in southern Vermont alone during late May 1988. Most trees put out a new set of leaves, but not without strain. Refoliation depletes food reserves in the tree, causing stress and, more importantly, lessening the tree's ability to deal with other stressors. Couple thrips defoliation with drought, tapping, acid deposition (formerly referred to as acid rain), disease, and the disturbance-induced stress of recent logging, and trees may die from a single defoliation.

Weather systems that bring haze and mist can be quite acidic and also carry other pollutants that affect trees. Acid deposition may not kill trees; but it can cause stress. One of the current theories is that the acids from the atmosphere displace aluminum in the soil which, in turn, pollutes important nutrient exchange sites on tree roots. The tree cannot take up essential elements such as calcium, and this situation eventually leads to twig dieback in the crown. Couple this stress with others and dieback becomes more severe, in some instances to the point of death. Acid deposition did not kill the tree, but it may have been the stress that tipped the scales.

Chapter 5
Forest Management Is Controlling Disturbances

Logging, though one of the most important silvicultural tools available, causes stress in many ways. Three are worth highlighting: 1) by opening the canopy, changing light and temperature levels; 2) by disturbing and compacting forest soils when heavy equipment is used for extraction; and, the most serious, 3) by causing wounds in residual trees, disrupting important physiological processes, and leaving wounded trees susceptible to decay and disease-causing organisms. Other potentially negative effects from practices such as high-grading ("take the best and leave the rest") reflect poor silviculture but often do not cause any more stress than logging to improve the stand.

Stands with closed canopies (highly shaded conditions) that are opened by logging may experience a phenomenon known as *thinning shock*. Although it is not exactly clear why this happens, it may be that trees divert energy into crown growth to take advantage of openings at the expense of providing adequate nutrition to other parts of the tree. Increased temperatures, light, and wind speeds are factors as well.

Thinning shock by itself is not bad. Usually the stand recovers after a growing season or two. But couple thinning shock with other stresses, and dieback and mortality will occur. **Closed stands that have been defoliated recently, or are suffering from drought, should be opened only very gradually or not at all.** The way to avoid thinning shock is to thin lightly. Doing so, however, leads to the potential of increasing stress from the other factors, namely soil compaction and wounds due to more frequent stand entry. By eliminating air spaces and decreasing the rate of water percolation, soil compaction can be a very serious side effect of harvesting.

Usually, 10–16% of a harvest area is devoted to roads, *log-landings,* and skid trails. Experts say that, with proper planning and good felling techniques, these areas could be reduced by 40%. Not only does good planning lessen the ecological impacts of logging; it makes timber harvesting considerably easier and cheaper. For example, the best way to prevent soil compaction and root disturbances is to schedule harvesting during frozen-ground conditions.

In any harvesting prescription — short of *clear-cutting* — wounding of residual trees is inevitable. Even the most careful felling techniques will cause some crown breakage, and stem wounding on residual trees used as *bumpers* is difficult to avoid. Of greatest concern, however, is wounding of feeder roots that can represent most of a tree's root system. Feeder roots are near the soil surface and are extremely susceptible to damage, especially during the growing season and when soils are wet, since wet soils shear more easily.

Root and stem wounding has increased dramatically with the use of *skidders* and other mechanized harvesting equipment. High-tech equipment can compensate for a lack of user skill with power, but there is always a tendency to over-size equipment at the expense of maneuverability.

Anytime the outside protective layer of a tree is broken, it creates an opening called an infection court. In response to wounding, some trees have evolved a method of walling-off or isolating the injury. As noted earlier, sugar maple is a tree that is good at walling-off its wounds; that is why it can recover quickly from repeated tapping. It is also why badly scarred sugar maples with full, healthy crowns standing next to a skid trail are not an uncommon sight. These trees, however, will never recover from the *grade* loss, or damage to lumber values, due to injury. Most early-successional species do not have the same ability as sugar maple to wall-off infections. Stem injuries in pioneer stands almost always result in a tree's demise.

Silviculture is a process of weighing the benefits of a prescription against the negative effects of harvesting. In the epilogue to Daniels et al., *Principles of Silviculture*, the authors cite a German forester, Henrich Cotta, who in 1816 wrote the following in his *Advice on Silviculture*:

The good physician lets people die; the poor one kills them. With the same right one can say the good forester [landowner, logger, or other who makes decisions about forest use] *allows the most perfect forest to become less so; the poor one spoils them.*

Logging is a double-edged sword. Regardless of the objective — be it for wildlife habitat, recreation, timber production, or a combination of uses — logging is the principal means of affecting and controlling change in the forest. It also has the potential to do great damage, some of which may be irreversible. Fortunately, most of the practices that lessen the negative impacts of logging are common sense.

Among woodland owners there is often confusion about the terms silviculture, forest management, and timber management. The word "silviculture" encompasses the practices that a manager employs to affect aspects of forest management involving trees. The term "forest management" implies control of forest resources (not just timber) for intended purposes that may or may not include wood production. For instance, a valid forest management strategy is to create habitat for a diversity of wildlife. In this instance, the manager uses silvicultural practices to improve habitats. All other uses are subordinate to wildlife, but not necessarily excluded. The term "timber management" implies control of forest resources primarily for wood production, but, again, not necessarily to the exclusion of other benefits such as wildlife, recreation, or scenic beauty. Silviculture is to a forest manager what tools are to a builder. The intended outcome — better habitat, more valuable timber, healthier ecosystems, or a combination — is the product of silviculture, just as a house — whatever its size, shape, or purpose — is the product of a builder.

Chapter 6
Silviculture in the Northeast

Silviculture is described by foresters as the "art and science" of growing and tending forests for the production of wood and other benefits. The word literally means "tree culture." It comes from the Europeans who, a few hundred years ago, first developed many of the ideas we accept and use today. Most of the early work is attributable to the Germans.

In Europe silviculture emerged as a method to ensure adequate wood supplies from a dwindling land base. If not for this scarcity, it is doubtful silvicultural practices would have evolved as the basis for managing forests for timber and other values. After all, it was not until the seemingly endless tracts of timber in the United States began to run out shortly before the end of the nineteenth century that silviculture and forestry took hold in this country.

That silviculture is both an art and a science is very important. Art is the result of creative inspiration; it is an expression of feelings and emotions, appealing to a personal sense of aesthetics. Science, on the other hand, is viewed as objective and impartial, an interpretation of facts. Feelings and emotions should have nothing to do with scientific inquiry. So defining silviculture as art and science may be seen as a contradiction. And yet, there is no question but that silviculture is a creative expression of practices that have a scientific basis.

The Practice of Silviculture

Most of our forests are the product of many years of use and, often, of abuse. Extensive clearing for pasture and crops during settlement in the 1700s was followed by abandonment in the late 1800s. This was followed by a long period of regrowth during the first part of the twentieth century and, since then, almost a lifetime of high-grading. No one knows how much growth potential forests have lost over the years as a result of high-grad-ing, but it could be substantial. Fortunately, our forests are very resilient. After nearly two centuries of use and abuse, the amount of forest cover is only slightly less than in presettlement times in some parts of the Northeast. Foresters have come to the conclu-sion that diligence and hard work are necessary to *prevent* for-ests from growing here! Our climate and soils are ideally suited to growing forests, and, though many of the species we try to grow are difficult to culture, trees of one kind or another will eventually prevail.

Most silvicultural decisions revolve around questions of species composition and forest structure. The silvicultural systems dis-cussed later (see chapter 7, "Silvicultural Systems," beginning on page 44) fall into either of two categories based on forest struc-ture: *even-aged* or uneven-aged *management*. Due to past land use, extensive areas have trees that are all about the same age; these are called even-aged forests. It means the age separation between the oldest and youngest trees is negligible. Since most of the region's forests are even-aged, we tend to focus on prac-tices that improve and regenerate this structure; and our most valuable timber species are well suited to these practices. How-ever, a disadvantage of even-aged silviculture is that at the end of a current rotation, after a new crop of trees has started on the ground, the mature overstory is removed and a young stand is left in its place. To foresters looking to the future, this scene is appealing. To the public, however, it may appear messy and destructive. And for species associated with the conditions of older forests, the dramatic change in habitat can be devastating.

One of the problems of practicing silviculture, especially in hard-wood stands on good soils, is that most of our forests have only attained 50–70% of their age of maturity and thus less than half their potential value as logs. Trees may have reached a size the market will accept, but they are growing at a rate that will yield increasingly valuable returns. Figure 6.1 (page 36) shows how growth rate is related to stand age. Harvesting trees in the middle of the curve, thinking they are mature and have quit growing, is a common and tragic mistake. Harvesting these trees at this stage is the financial equivalent of selling a certificate of deposit be-fore it has matured; you pay a substantial penalty for doing so. For this reason, much of the harvesting done to fulfill a silvicul-

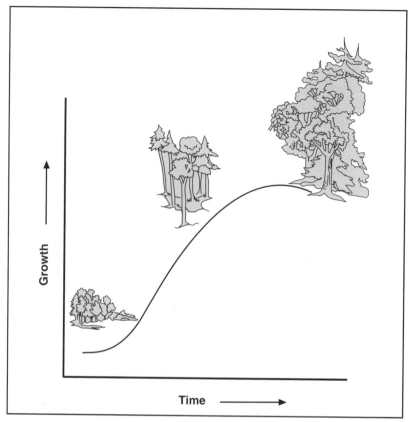

Figure 6.1 Growth rate is related to stand age.

tural prescription in even-aged forests is intended to improve growing conditions for the trees left behind. This is called an *intermediate treatment* and yields relatively small volumes of low grade sawtimber, pulpwood, and fuelwood (also see "Intermediate Treatments," beginning on page 38).

Intermediate treatments are often difficult to sell to timber buyers because product values are low and the need to protect the residual stand from logging damage is high. Timber buyers and loggers are reluctant to operate under these circumstances unless some highly marketable timber is available. This is a common concession managers make, but if overdone the concessions may lead to high-grading, loss of future timber values, and damage to the site.

Introduction to Forest Ecology and Silviculture

Silviculture is a process of managing trade-offs. Without strong markets for wood, there are no *stumpage* buyers; without buyers there is no work for loggers; without loggers there are few opportunities to practice silviculture; without silviculture there is a less desirable product mix. The trade-off involves designing treatments that are economically feasible while enhancing future timber values.

Unfortunately, many private, nonindustrial forest owners do not have silviculture in mind when they harvest timber. Some owners harvest all merchantable timber just before selling the land. But despite their need for money, there are no valid silvicultural terms to describe what is, in fact, a liquidation of assets also known as *liquidation cutting*. This is not to say that an owner has no right to extract as much value from the land as possible. Rather, it is important to realize that a liquidation of timber is not silviculture.

The Purpose of Silviculture

Almost any planned activity with the purpose of changing conditions to improve forest value is silviculture. Whether it is to improve scenic vistas, create or enhance habitat for specific wildlife, or increase the investment potential of timber, if there is an objective that involves manipulation of the vegetation to achieve a positive outcome, it is properly characterized as silviculture.

Most often silvicultural practices are employed to improve conditions of trees in forests for investment purposes. In this context — growing trees for wood production — silviculture encompasses a wide range of practices intended to reproduce forest stands or to increase the growth rate, vigor, and value of trees. The unit of silviculture, or the level at which treatments are designed, is the stand. A prescription is a course of action to effect change in a forest stand. When timber production is the goal, the change is usually one of two alternatives: 1) to improve growing conditions for the trees left on the site after the prescription is carried out; or 2) to get a new stand of trees started. Most silvicultural practices that involve cutting are intermediate treatments intended to improve the existing stand or *regeneration* treatments intended to get a new stand started (figure 6.2).

There are many other silvicultural treatments, such as fertilization and pruning, but this book focuses on practices that are easily implemented by cutting trees.

Regeneration or Improvement?

Good forest managers are always thinking about the future. When the prescription calls for improving the existing stand, the future is more near-term, or at least within the rotation of the current stand. The rotation is the age to which trees in a stand will be allowed to grow until final harvest.

When the prescription calls for regeneration, the future is more distant. Here the forester is trying to create conditions favorable for the next forest. It takes time: the forester who prescribes a regeneration treatment may live to supervise the first *precommercial thinning*, but after that, following through on the prescription becomes the responsibility of another forester.

Intermediate Treatments

Practices to improve stands are called intermediate treatments because they take place after a stand is started but before final harvest. *Timber stand improvement*, or *T.S.I.*, is a common term to describe intermediate treatments to improve the existing stand.

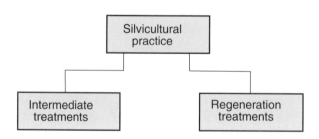

Figure 6.2 Most silvicultural practices that involve cutting are intended to improve the existing stand (intermediate treatments) or to get a new stand started (regeneration treatments).

The two main reasons for T.S.I. are 1) to adjust species composition to suit the site or meet anticipated markets; and 2) to thin the trees, favoring selected *crop trees* by increasing the rate at which they grow in diameter.

If the goal is to maximize lumber yields from timber, periodic thinning is essential. Thinning reduces the number of stems and gives space, soil nutrients, and moisture to residual trees. The purpose is to maintain or increase the rate of diameter growth of residual trees. But there is a sacrifice: fewer trees are available for harvest at a later date, so thinning is a choice between growing many trees of relatively small diameter or fewer trees of larger diameter. Volume growth rate per tree or per acre is controlled by tree density (figure 6.3). Well-spaced trees optimize volume growth on individual trees, but without sacrificing — to an unreasonable extent — total volume growth per acre.

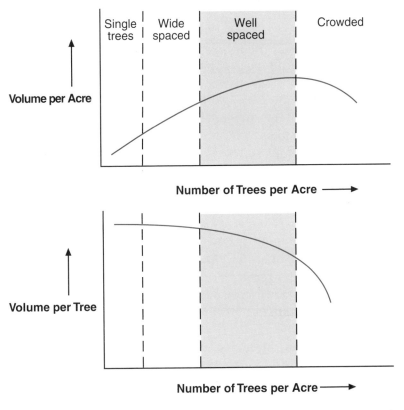

Figure 6.3 Volume growth rate per tree or per acre is controlled by tree density.

The optimum is usually difficult to achieve, and it changes as the stand grows. Adding to the complexity of deciding when to thin and by how much is a need to assess the impact of damage to the residual stand caused by felling and skidding. Injuries to stems and roots from logging are more damaging to future log grades and tree growth as the stand gets older. Sometimes the benefits of thinning to tree growth are wiped out by logging damage.

Stocking and Density

Stocking is an expression of the *density* of trees in a stand. More specifically, it is a relationship between the optimum number of stems per acre and the average diameter of trees in the stand. Most intermediate silvicultural practices are intended to control stocking, managing the trade-off between optimizing diameter growth of individual trees and optimizing volume growth per acre (figure 6.3, page 39).

Two measures used to estimate stocking are 1) the number of trees per acre, and 2) the total cross-sectional area of tree stems at d.b.h., or 4.5 feet above ground. The sum of the cross-sectional area of tree stems, measured at d.b.h. and expressed as square-feet per acre, is known as *basal area* (figure 6.4). The greater a stand's basal area, the greater the proportion of an acre that is occupied by tree stems.

Basal area is one measure of density. But to say the basal area of a particular stand is 100 square feet per acre does not say anything about the number of trees or their size. It could be more than 500, 6-inch trees or about 130, 12-inch trees (figure 6.5). Both basal area and number of trees per acre are necessary to estimate stocking. *Mean stand diameter* is a relationship between the number of trees per acre and basal area. As trees grow in diameter and the stand grows in volume, basal area is concentrated on fewer stems.

In undisturbed natural stands, basal area increases as trees grow in diameter, and the rate of diameter growth is controlled by stem density. As some trees die from natural causes — wind, fire, insects, and disease — surviving trees grow to take up the space. In older stands basal area growth occurs on fewer and fewer trees. In very old stands, diameter growth has almost

Introduction to Forest Ecology and Silviculture

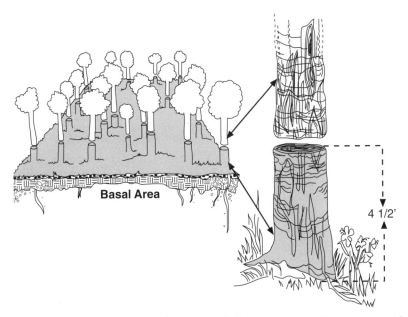

Figure 6.4 Basal area is the sum of the cross-sectional area of tree stems measured at d.b.h.

Mean Stand Diameter in Inches

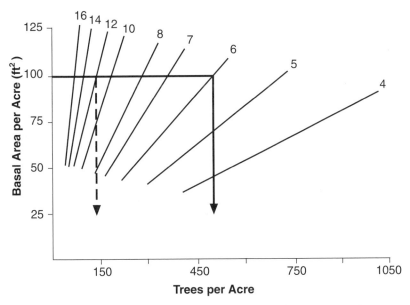

Figure 6.5 Mean stand diameter is a relationship between the number of trees per acre and basal area.

stopped and a few very large trees account for most of the basal area.

The timing of thinning to increase the rate of diameter growth is based on stocking. Optimum stocking is the point where both volume growth per tree and volume growth per acre are maximized. Research foresters have identified optimum stocking levels for different forest types.

In figure 6.6, the A-, B-, and C-lines are stocking thresholds for even-aged stands. The thresholds are defined as follows: above the A-line is overstocked, or too many trees for good basal area growth on individual trees. Below the C-line, the stand is said to be understocked, a situation where basal area growth per stem is high but there are too few trees to optimize the rate of basal area growth per acre. The B-line is optimum stocking; it corresponds to "well spaced" in figure 6.3 (page 39). Between the B-line and the A-line, stocking is adequate, but the closer stocking is to the A-line, the more the stand needs thinning. Below the B-line, the stand is too thin but it should reach optimum within

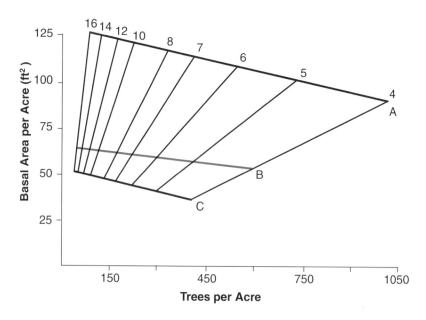

Figure 6.6 Stocking chart for northern hardwoods in New England

ten years. The position of the lines is based on research and varies by forest type. Estimates of basal area and numbers of trees per acre are necessary to use stocking charts. Figure 6.6 is a stocking chart for northern hardwoods in New England. Some additional common charts are included in appendix B (page 72).

After a thinning, residual trees are free to grow for ten to twenty years or longer before another thinning is necessary. Since thinning is a process of removing trees, basal area growth occurs on fewer trees after thinning but at a faster rate than would occur naturally.

Most of the concepts that relate to stocking and thinning assume that stands are even-aged. In a well-managed stand, thinning may happen two to five times or more during a single rotation. As the stand gets older, the temptation to harvest crop trees is hard to resist. "Making space for smaller trees" is often cited as the reason for taking the best trees from the stand; that is, removal of a large-diameter tree next to a smaller one will allow the smaller to grow. **But actually the larger tree should be left to grow, as it has proven to be a better competitor than a nearby smaller tree of the same age.** Deciding when to harvest a particular tree, however, is a question of health, condition, and maturity, not necessarily size.

Late in a stand's rotation, during the final thinning cycles, foresters start to think about regeneration — the next stand. Depending on method of regeneration, the emphasis shifts from concern for crop trees to getting a new crop started. The thinning prescription may change to improve conditions for reproduction of the selected species. When it is near time to regenerate, the existing stand is used, to varying degrees, for cover and protection of seedlings while they are becoming established in the understory.

Chapter 7
Silvicultural Systems

A silvicultural system is a program of management defined by 1) the methods used to regenerate the stand, and 2) the structure of the managed stand; that is, whether it is even-aged or un-even-aged. The systems vary in the severity of disturbance used to achieve regeneration: from the most severe — clear-cutting whole stands; to the least — removing scattered, individual trees as they mature. The most severe treatments, such as clear-cutting, favor early-successional species such as aspen and white birch; while the less severe treatments favor late-successional species such as sugar maple, beech, and hemlock.

The regeneration treatments are discussed below in decreasing order of disturbance (most to least), beginning with those that result in formation of even-aged stands.

Even-Aged Methods

Clear-Cutting

One of the most publicly maligned and misunderstood regeneration treatments is clear-cutting (figure 7.1). It is a legitimate silvicultural practice, under certain circumstances. Since clear-cutting is also the most efficient (least costly) method of removing wood from forests, there are instances where economic efficiency is confused with silviculture. Impacts on aesthetics, water quality, and recreation are often cited as reasons why clear-cutting should be abolished. But when done properly, clear-cutting is highly beneficial, especially to many species of wildlife; and negative water quality and visual impacts are usually easy to minimize.

Clear-cutting is used to regenerate species, such as aspen and white birch, that are adapted to catastrophic events such as hur-

Figure 7.1 Clear-cutting

ricanes or fire. To be most effective, the clear-cut should duplicate the light and temperature regimes of an open field, the conditions under which early successional species do well. Depending on site, slope, and aspect, effective clear-cuts range in size
from a few acres to a few hundred acres.

Clear-cutting is the only successful method for reproducing aspen or popple, a species that benefits many types of wildlife. It
has light, wind-borne seeds that quickly colonize bare soil openings. The roots of felled aspen stems also produce root-suckers
when ground temperatures are high. Aspen is mostly regener-
ated by root suckering in forests. Seed is usually not important.

The distinction between clear-cutting as a silvicultural tool and
logging practices that remove all marketable timber is important. The objective of clear-cutting is reproduction of a new forest. When the sole purpose of logging is to convert trees to wood
products, even though the outcome may look similar, it is not
clear-cutting. The objective of liquidation cutting is to convert
trees to cash, and usually what follows is a quick sale of the land
and whatever is left standing. The difference between liquidation cutting and clear-cutting is that one is done for purely eco-

nomic reasons, the other to effect change in forests for beneficial purposes.

Seed Tree

A modification of clear-cutting hardly ever used in the Northeast is the *seed tree* method (figure 7.2). Trees are left widely scattered on the site following harvest to provide seeds for regeneration. Timing is crucial to ensure that the harvest coincides with a good seed year. Failed seed tree regeneration treatments, when reproduction is delayed or the species mix is wrong, are usually replanted with seedlings. Seed tree methods are used mostly in the southeastern U.S.

Shelterwood

In the *shelterwood* method, a partial overstory is left to provide shelter for seedling establishment (figure 7.3). Depending on the degree of cover, the shelterwood method favors mid- and late-successional species.

The shelterwood usually has two phases: the first, called a seed cutting, to secure adequate regeneration; and the second, called a removal cutting, to harvest the overstory and leave the site to a young, developing stand. Occasionally, a shelterwood has three treatments. The first, called a preparatory cutting, removes species that may be a problem in the new stand and stimulates seeding of desired species. The second and final cuttings are as before.

Shelterwood is probably the most widely used regeneration system in the region where the objective is to get started a new, even-aged stand of trees — hardwood or softwood. It is much more acceptable to the public than clear-cutting, since a protective overstory remains in place after the preparatory and seed cuttings. Moreover, scheduling of the final harvest can be delayed until markets are favorable or other work in nearby stands needs attention. Good record keeping, however, is essential to achieve success with the shelterwood methods. If the final harvest is done before regeneration is adequately established, many seedlings may be lost due to exposure (mostly due to drying

Figure 7.2 Seed tree cutting

Figure 7.3 Shelterwood

conditions caused by increased wind speeds and higher temperatures). If the final harvest is delayed too long, however, the new sapling stand may be severely damaged by equipment when the overstory is finally removed.

An interesting variation of the shelterwood method used by some foresters in the region is known as *deferment cutting.* The seed cutting is a little heavier, leaving more high-risk and cull trees as seed sources, and the new stand is left to mature with the sheltering overstory in place. The removal cutting is timed to coincide with the first thinning, or even later in the rotation.

Depending on the degree of concern for regeneration and the manager's perception of how long it might take to establish an acceptable mix of species, the shelterwood method can take the form of a series of thinnings late in the rotation. The result is a higher proportion of late-successional species established over a longer period of time. Under these circumstances, the shelterwood starts to look more like the uneven-aged methods discussed below.

Uneven-Aged Methods

When trees in a stand have a range of ages from young to old, the stand is called uneven-aged. A balanced uneven-aged stand looks like an undisturbed, mature forest (figures 7.4 and 7.5). As older individuals die, seedlings of late-successional species — that have been idling along in the understory — begin to fill in the gaps caused by *windthrow,* disease, insects, fire, or other disturbances. This results in small pockets of roughly same-aged trees, but for the stand as a whole, the structure is uneven-aged (figure 7.5). Undisturbed old-growth forests typically exhibit a balanced, uneven-aged distribution of tree diameters, and all age classes are represented — from seedlings to old-growth (figure 7.4).

Uneven-aged methods were developed to try to duplicate the natural replacement of trees in old-growth forests.

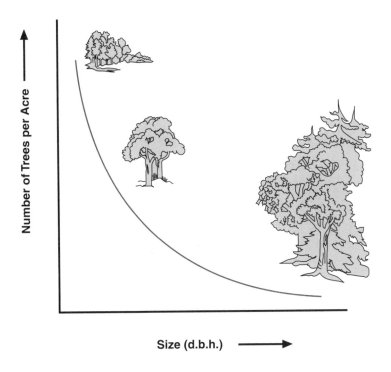

Figure 7.4 Old-growth stands are usually composed of pockets of same-aged trees that fill in gaps caused by disturbances.

Figure 7.5 Undisturbed old-growth forest typically exhibits a balanced, uneven-aged distribution of tree diameters and all age classes.

Group Selection

As its name implies, the *group selection* method removes small groups of trees (figure 7.6). But because the opening is small, the gap is shaded for most of the day, and climax, shade tolerant species will eventually fill in. If the groups are larger and there is a source of seed nearby, mid-successional species such as white ash and yellow birch (on northern hardwood sites) are favored as well.

The idea here is to identify areas ready for regeneration. But before groups are identified, one must assess site quality and tree size, health, and maturity. Existing regeneration or potential sources of seeds are also identified. The goal of uneven-aged methods is to maintain a wide variety of tree ages and diameters. For the first few cutting cycles (when the stand is twenty to sixty years old), most emphasis is placed on achieving a favorable distribution of tree diameters. This ultimately results in a continuous succession of mature trees. It may take forty years or longer to convert an even-aged northern hardwood or central hardwood stand to an uneven-aged (or at least uneven-structured) stand.

Single-Tree Selection

The method of regenerating forest stands that can potentially cause the least disturbance is single-tree selection (figure 7.7). Only fully mature trees are harvested, but for the first few cutting cycles some trees are removed from almost all diameter classes (including high-risk, poorly formed, and *cull* trees) to create a more even succession of mature trees. The single-tree method is often used in conjunction with group selection, especially where there is a better opportunity for regeneration if more than one tree is removed. Single-tree selection methods should not be confused with other similar sounding terms, such as *selective harvest*, which, more often than not, leads to high-grading (also see "Myth 2," page 63).

Successful single-tree and group selection methods are difficult. They require fairly frequent stand entries (usually every five to ten years) and extreme care by *fallers* and skidders. Removing a mature tree without damaging nearby saplings and poles is a

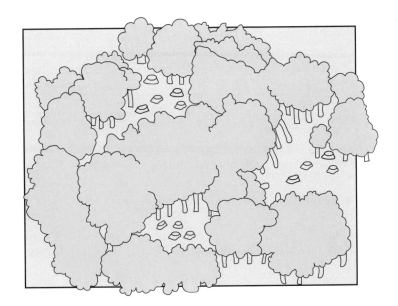

Figure 7.6 Group selection

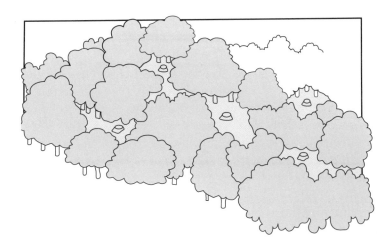

Figure 7.7 Single-tree selection

challenge for even the most skilled logger. Selection methods also require more record keeping than even-aged methods and are usually more costly due to lower harvest volumes and the higher degree of care that is necessary.

Selection methods have become increasingly popular in recent years, especially on productive sites, because — if the methods are done correctly — soil disturbance is minimal and the visual effects of logging are less evident. Since many owners do not rely on income from their forests, the inefficiencies of harvesting single trees or small groups is not a big concern. A drawback to single-tree selection is the need for frequent stand entries. If improperly planned and executed, these entries can cause extensive damage to the crowns, stems, and root systems of residual trees. For these reasons, selection methods are usually most successful when implemented during frozen-ground conditions.

Chapter 8
Developing Silvicultural Prescriptions

Silviculture is an attempt to control disturbances in forest stands to achieve favorable results — such as faster timber yields, higher stem quality, acceptable regeneration, and improved habitat. In fact, the entire basis for silviculture hinges on our ability to predict the outcome of a treatment. Predicting an outcome implies objectives. Without clear objectives there is no silviculture; it is just cutting trees.

Forest management objectives and silvicultural objectives are not the same thing. The management objective is a guiding principle — a statement about what is important to the owner. The silvicultural objective is specific to each stand, but recognizes and supports the overarching management objective by specifying treatments for what is desired. For example, a forest owner may express a management objective to "protect habitat for songbird species that use climax forests." A silvicultural objective for this same owner might be, "Manage for late-successional northern hardwoods on long rotations using uneven-aged regeneration methods and a ten-year cutting cycle."

Many factors have a bearing on the outcome of a prescription, or whether or not it is successful. These include capabilities of the stand, owner objectives, market demands, and skills of the forester and logger (figure 8.1, page 54). Market demands, in particular, seem to dictate prescriptions; when market forces are strong, the prescription can suffer. This leads to "market-driven silviculture" which is most unfortunate, since markets change. Many foresters in the region still remember a time when red oak and yellow birch were considered weed species. Today they are two of our most valuable species, but who can say if this will be true fifty years from now?

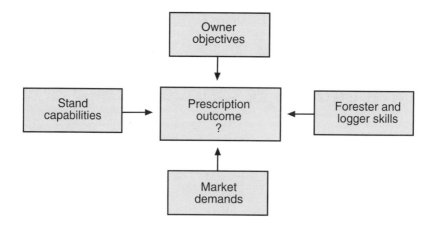

Figure 8.1 Many factors in combination determine the success or failure of a silvicultural prescription.

A Scenario

Up to this point it may seem as though the most complicated part of developing a silvicultural prescription is choosing which trees to cut. The fact is, though, a lot of thinking goes into a prescription. Depending on the owner's management objectives, the property is inventoried and mapped (usually by a qualified forester); intensively if the objectives are strictly to harvest timber, less intensively if the objectives are otherwise, such as to enhance recreational opportunities. The inventories are used to develop silvicultural prescriptions which, in turn, guide which trees are to be marked for cutting. Figure 8.2 illustrates that the woodland owner's objectives determine the amount of stand mapping and inventory that are necessary, and these objectives also provide guidance for stand-by-stand prescriptions.

Assume the forest owner in this scenario has "long-term timber investment" as her primary forest management objective. The following scenario is based on "Stand 2," a 65-acre stand of northern hardwoods. Figure 8.3 shows a page from the management plan for Stand 2, describing stand structure, timber volumes, site index, and other information.

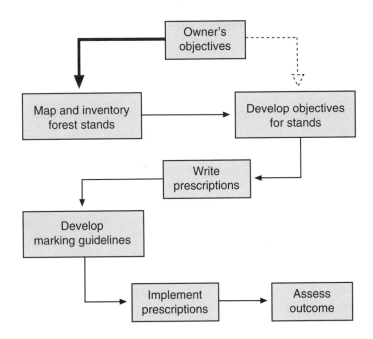

Figure 8.2 The woodland owner's objectives determine the amount of stand mapping and inventory and provide guidance for stand-by-stand prescriptions. Intensive inventory and detailed mapping may not be necessary.

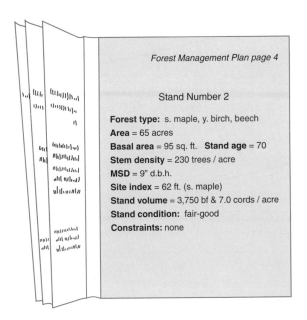

Forest Management Plan page 4

Stand Number 2

Forest type: s. maple, y. birch, beech
Area = 65 acres
Basal area = 95 sq. ft.   **Stand age** = 70
Stem density = 230 trees / acre
MSD = 9" d.b.h.
Site index = 62 ft. (s. maple)
Stand volume = 3,750 bf & 7.0 cords / acre
Stand condition: fair-good
Constraints: none

Figure 8.3 A page from the management plan for Stand 2

Given the overall management objective of "long-term investment," the current condition of the stand and its future growth potential, and site quality, a reasonable silvicultural objective might be to

> **Produce high-quality sawtimber in an even-aged stand using the shelterwood system and a 125-year rotation.**

The stand data for basal area and numbers of trees per acre is plotted on the stocking chart for northern hardwoods (figure 8.4). The B-line is the recommended stocking level to maximize both individual tree volumes and volume growth per acre. In figure 8.4, the before-treatment stand structure is shown by the intersection of basal area and number of trees per acre, and the after-treatment structure is indicated by the dashed lines. Subsequent thinning is scheduled when basal area growth increases

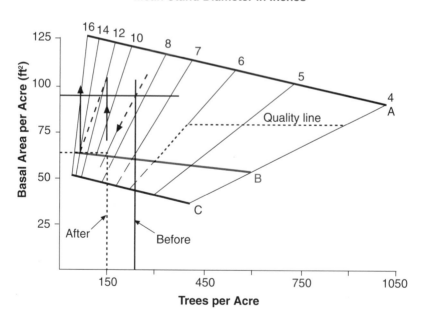

Figure 8.4 A stocking chart for northern hardwoods showing the before-treatment stand structure and after-treatment structure

the mean stand diameter to about 12 inches. At this point, the stand is ready to be thinned back to the B-line once again. The chart is used to estimate after-treatment stocking, in this case about 150 crop trees and 65 square feet of basal area. Stocking charts such as this are used to maximize growth rates of selected trees without overly sacrificing volume grown per acre.

This information is used to develop the following prescription:

Select crop trees and thin the stand by reducing the basal area to 65 square feet per acre.

The prescription is used to prepare marking guidelines, the rules to ensure that the prescription and the objective are fulfilled. The marking guidelines are used by the forester to help make tree-to-tree decisions about what stays and what goes.

In this scenario, the directions for marking the stand are as follows:

Remove poorly formed, diseased, cull, high-risk, and low-value stems. Leave sugar maple, yellow birch, and healthy beech. Avoid marking in areas where damage to the residual stand is apt to be high. The marked-tally of trees for removal should equal 80 to 120 stems per acre.

The yield from this prescription is marginal: 1,100 board feet of low-quality sawlogs and 4.5 cords of fuelwood per acre (71.5 MBF and 292 cords for the 65 acres). If done properly, however, the outcome of a prescription such as this is a well-stocked stand of healthy, well-formed trees with crowns opened-up on at least two sides. Residual stand damage is acceptably low (less than 5% of the residual trees show injuries), and site disturbance from harvesting is minimal. Within three to five growing seasons, trees in the stand should show a noticeable increase in diameter growth. Successive thinnings — perhaps as many as three before the final harvest — should yield increasingly higher-quality logs.

The treatment proposed in this scenario results in a small, but *operable*, timber harvesting opportunity. There are many instances where intermediate treatments do not yield products.

The question is often posed, "Who pays for silviculture?" The answer is, "The person who owns the woodland, not the stumpage buyer or the logger." Woodland owners and their foresters are very willing to negotiate stumpage rates to get a good job. After all, a poor job today destroys the prospects for substantial gains in the future. Good silviculture is partially the logger's responsibility, but it is the person who owns the land — the investor — who must pay the cost of good silviculture.

Practicing Silviculture in Stands Susceptible to Ice Damage

It is safe to say that every acre of forest in the region has been affected by ice storms. The most damaging events happen during midwinter when very cold air is in place and a strong, wet low-pressure system rides up and over the cold air. As rain falls it freezes on contact with cold surfaces. Usually the conditions for ice buildup do not last for more than a few hours, but occasionally weather systems stall and freezing rain can cause ice formation in trees to the extent that branches break and unbalanced trees are uprooted. In mountainous areas, near ridgelines, usually on east- and northeast-facing slopes, forest stands are perennially affected by ice, so much so that these areas are characterized as "ice-affected ecosystems." These stands are easy to spot by the presence of broken limbs and generally misshapen crowns. Planning intensive timber management in these areas is pointless because the periodic damage from ice can slow growth rate and have a substantially negative effect on timber values.

In other areas susceptible to temperature inversions that can lead to ice damage, areas where the surrounding topography is higher, creating a bowl effect that serves as a basin for cold air, stands should be maintained at higher stem densities than would otherwise be prescribed. Higher stem densities allow ice-laden branches to scaffold into one another, greatly increasing their strength, so that branches can withstand much higher ice loading (figure 8.5). Employing such practices means more frequent,

Figure 8.5 Avoid heavy thinning in stands that are prone to ice damage, so that ice-laden branches can scaffold into one another.

lighter thinnings and some loss of diameter growth. But in the event of a severe ice storm, stand damage will be appreciably less when tree crowns are close enough to rely on each other for support. Just so long as the main crown is intact, broken branches — even to the extent of 50% of the crown — will not necessarily lead to tree mortality.

Trees have a higher tolerance to damage in winter because humidity is low, causing exposed wood to dry rapidly, and there are fewer disease-causing organisms in the air. Light to moderate crown thinning from ice damage may even stimulate diameter growth. When crown damage exceeds 50%, however, the tree is a candidate for removal unless there are good reasons to keep it in the stand. When crown loss reaches 75% or more, the tree is unlikely to recover and should be removed within the next few growing seasons to avoid grade losses from wood stain and decay.

Chapter 9
Combining Timber Goals with Other Resource Values

Silviculture is more than growing trees for the sawmill. Many silvicultural practices are easily adapted to other, non-timber, benefits. In most instances it is very easy to alter a prescription to provide resource benefits in addition to sawtimber. Some timber potential is lost, but it is usually minimal. In addition, most forest owners like to maintain their forests for wildlife and other values. For example, a forest owner who sees wildlife as an important resource may want to preserve aspen clones. Old aspen clones can be easily reproduced by removing the trees in the clone and opening the area to sunlight (a mini-clear-cut). Almost any prescription — improvement or regeneration — is easily altered to treat old aspen clones.

In the north, *strip cuts* in spruce-fir stands are an excellent way to regenerate these species while preserving the structure of the *yard* for wintering deer. The goal is to keep the strips narrow, usually a tree-height or less, to reduce snow accumulation and allow deer to travel between strips. Windthrow is a common hazard in these stands, so strips are usually oriented east-west, or northeast-southwest, depending on the surrounding landform and windthrow risk. This orientation also maximizes shading of the cut strips, lowering soil temperatures and improving conditions for spruce-fir establishment.

Many years ago the Germans developed a method to grow sawtimber on long rotations and fuelwood for their stoves in the same stands. The method, called "Coppice with Standards," employs a 20- to 40-year rotation for understory trees grown from stump sprouts, and an 80- to 120-year (or longer) rotation for a fairly sparse overstory of *standards* — trees of seed origin. In this way four to six fuelwood rotations are conducted during the course of a single sawtimber rotation.

Today, most managers use silvicultural prescriptions to accomplish more than just growing timber. Often it is the opportunity to achieve some other objective, or combination of objectives, that serves as justification for silviculture. The trick is to recognize those opportunities and to design practices that can optimize both timber and non-timber benefits. For example, the sidebar lists guidelines for silvicultural practices that are beneficial to wildlife.

Sidebar
**Guidelines for Silvicultural Practices
That Are Beneficial to Wildlife**

With just a little extra planning, timber management practices can be highly beneficial to wildlife. Here are some things to consider:

Timing — The greatest disturbance to breeding bird populations, especially to migrant birds that rear young in the North, is logging from May to July. If you suspect harvesting may disturb breeding, ask someone from your Department of Fish and Wildlife.

Hard mast — Beech, oak, and hickory nuts are a high-energy food source for many species. Leaving even a few mature trees per acre, especially trees that look like good seed producers, is beneficial.

Buffers — Buffer strips along streams help regulate stream temperatures and protect breeding fish populations.

Cavities — Leave some trees with cavities for wildlife species that use them for shelter.

Landings — Seed in *log landings* to provide food and cover for many woodland wildlife species.

Diversity — Good stand structural diversity and species diversity will support more varied populations of wildlife than uniform, single-species stands.

The big picture — improving woodland habitats is often a matter of providing the missing piece of the puzzle in a larger landscape. Look to surrounding woodlands and use the stand you are working in as an opportunity to provide an element of habitat that may be missing over a much larger area. This, more than anything else, is the secret to managing woodland habitats for wildlife.

Source: Adapted from Toni McLellan, Wildlife Biologist, United States Forest Service, Durham, New Hampshire

---

Chapter 10
Common Myths about Silviculture

Some of the most common misconceptions about silviculture arise because of our difficulty in planning for events that extend beyond our lifetime. We make decisions about a resource that takes nearly two lifetimes to mature, yet by nature we are impatient. For these reasons and others, myths about silviculture abound. Here are some of the most misleading.

Myth 1: Big trees are more mature than little trees.
Reality: Possibly the most common mistake of people who are new to forestry and logging is thinking that the bigger trees in an even-aged stand are older and more mature than the smaller trees. The fact is that the bigger trees are more vigorous and should be left to grow while nearby smaller trees should either be 1) removed from the stand if they are an early-successional species or a mid-successional species that has been suppressed; or 2) retained only if a) they are late-successional species with excellent form and there is a good chance they will reach the overstory before the end of the rotation; b) they will be carried to maturity if the stand is to be converted to an uneven-aged structure; or c) they contribute to other management goals, such as improving habitat.

Sometimes harvesting is done according to diameter limits: only trees exceeding a certain diameter (d.b.h.) are harvested. If the limit is low, as when all trees 12 inches d.b.h. and larger are harvested, 80% or more of the stand's merchantable volume may end up on the landing. This is an extreme case of "making space for the smaller trees" and is a silvicultural catastrophe. Diameter-limit cutting is not a valid guideline for even-aged stands in the Northeast. It is high-grading masquerading as silviculture.

Myth 2: A valid silvicultural treatment is the selective harvest.

Reality: Somehow the term "selective harvest" has become popular in the region. Used to describe a partial harvest (something less than a clear-cut!), it implies *only* that someone has selected the trees for harvest. It is often confused with those terms used to describe uneven-aged methods, but **"selective harvest" has no silvicultural meaning and should not be confused with group and single-tree selection methods.** This is another example of high-grading masquerading as silviculture. (Also see "Single-Tree Selection," beginning on page 50.)

Myth 3: If it looks good when you are through, it is good silviculture.

Reality: Good silviculture does not always look good, especially in the first year or so after a treatment. Yet many woodland owners equate a good job with good visual appeal. This is not to say that aesthetics are not important. Nonetheless, a prescription should not be compromised just to make it look good. There are many ways to protect the visual appeal of landscapes during and after harvesting, such as leaving *buffer strips* near trails and roads. (See, for example, Beattie, Thompson, and Levine).

Myth 4: A marked stand is a silvicultural prescription waiting to happen.

Reality: Anyone with paint can mark trees for harvest. Just because a stand is marked does not mean the person who marked it had a silvicultural objective in mind. Most of the time it is helpful to ask the forester about the marking guidelines and the prescription. Note that once the stand is marked, paint of a similar color is banned from the site and cause for termination of a logging contract.

Myth 5: Silvicultural treatments make stands more productive.

Reality: The long-term effect of repeated timber harvests often is a gradual lowering of timber growth productivity. Heavy equipment operating on some sites at certain times of the year can be devastating. However, the practice of silviculture can lessen the impacts of harvesting on sites, and is usually intended to concentrate wood production on fewer stems. Silviculture can increase the productivity of stands for lumber production, but *biomass* productivity is usually less as a result of repeated timber harvests.

Myth 6: If a stand is suffering from stress, thin it out.

Reality: We used to think that thinning a stand was a great way to help trees recover from insect and disease outbreaks or other stresses. But harvesting is another stress that, compounded with other stresses, can lead to crown dieback and even mortality. If a stand is under excessive stress, the best thing to do is leave it alone until it shows signs of recovery.

Chapter 11
Working with Foresters and Loggers

Foresters and loggers influence the outcome of silvicultural treatments in a great many ways. Good communication with those who develop and implement prescriptions is absolutely essential. Communication failures lead to silvicultural failures. Loggers should ask foresters questions in order to have a clear understanding of the objectives behind treatments that loggers are asked to implement. Knowing if a prescription is intended to secure regeneration or to improve the growth and value of a residual stand will have a tremendous bearing on the best way to operate. And remember: it is the logger's responsibility to protect the site from the potentially negative impacts of harvesting. **However, the success or failure of a prescription is the combined responsibility of the forester, the landowner, and the logger.**

Favorable markets can make logging more profitable, but a decision made solely on the basis of current markets is not sound silviculture. Resisting the lure of strong product markets is often the most difficult aspect of silvicultural practice for loggers and foresters, especially when loggers and foresters represent wood-using companies. When a logger's financial well-being is on the line, it is difficult to pass up current income opportunities in anticipation of future gains. "After all," many stumpage buyers reason, "if I don't take the timber today, someone else will take it tomorrow." No matter how true this statement may be, we all have a responsibility to ensure that future forest values are protected. Sometimes this may mean educating the forest owner, and sometimes just walking away. Aldo Leopold, a well-known forester, wildlife biologist, and writer, said it best:

A system of conservation based solely on economic self-interest is hopelessly lopsided. It tends to ignore, and thus eventually eliminate, many elements in the land community that lack [current] *commercial value, but that are (as far as we know) essential to its healthy functioning. . . . An ethical obligation on the part of the private owner* [and foresters and loggers] *is the only remedy for these situations.*

Chapter 12
Ecosystem Management

The practice of forest management evolved in Europe from a need to sustain wood production on a decreasing land base. As land was converted to agricultural uses or occupied by human settlement, less land was available for timber production. Since "scarcity is the mother of invention," the Europeans developed techniques to improve tree growth and wood yield. These practices were imported to this country in fairly recent times, not more than 150 years ago. Forests are one of the few natural resources we have that are renewable. In other words, we can use wood fiber, and eventually trees regenerate and harvestable wood is available again and again. Silvicultural practices imported from Europe have served us well in the years following a period of exploitation — throughout most of the nineteenth century. But in just the past 10 years, increased recognition of the interrelatedness of forest organisms and the potentially negative impacts of timber management practices on non-timber values have resulted in new ideas about how we manage and use forests. These ideas are commonly referred to as *ecosystem management*, but they not are fully accepted by all professionals, some of whom believe that traditional forest management methods — when correctly employed — protect and improve forest ecosystems.

Silviculture encompasses practices that are primarily intended to improve forests for human use, and traditional forest management methods emphasize timber growth and extraction. Foresters believe silvicultural practices, when employed with a thorough understanding of forest ecology, are consistent with maintaining healthy ecosystems, but it is on the definition of "healthy" where opinions diverge.

The difference of opinion centers on what we manage woodlands for, and who benefits. Proponents of ecosystem approaches say the emphasis should be placed on *forest health*. In other words, we should concern ourselves with future forests and future generations. Managers, according to this view, have a responsibility to control disturbances so as to lessen the negative impacts of logging, and to use practices that mimic nature. Human use, above all, should not be detrimental to other forest organisms — from songbirds to microscopic soil bacteria. Ecosystem management, or ecosystem approaches to management, suggests that responsibility for understanding the complexities of forests rests with those who use forests. In addition, our management practices, while attempting to mimic nature, must change as we learn; when we don't know, or can't reasonably predict, the outcome of a practice, we should err on the side of inaction rather than risk subtle but irreversible impacts when we proceed without knowing.

The days of "cookbook" silviculture are nearly over, as are universal prescriptions aimed solely at increasing timber volume and value, to the exclusion of less tangible, but no less important, ecosystem values. Under the guidelines of managing whole ecosystems, forests will still provide timber, but timber production will be subordinate to the long-term health and *sustainability* of forests. Disturbance — sometimes catastrophic changes brought on by storms, insects, disease, and fire — is the rule in the Northeast rather than the exception. Ecosystem management means designing disturbances that mimic natural processes in a way that allows us to use the resource while protecting the integrity of the landscape.

Figure 12.1 suggests that an ecosystem can be defined at many different levels, from the lichens growing on the bark of a single tree to entire landscapes, continents, and — ultimately — the earth. The hierarchical effect is what makes ecosystems so complex, enchanting, and challenging.

Practicing ecosystem management will eventually challenge our interpretations of private property rights because ecosystems are nested one within the other and do not follow geographical, cultural, or political boundaries. Some migratory songbirds, for example, winter in tropical regions of the Caribbean and the Americas but use northeastern forests during the spring and early summer for nesting. Why they migrate, we don't know. But it may have something to do with the rich supply of proteins avail-

Figure 12.1 An ecosystem can be defined at many different levels.

Source: Adapted with permission of publisher from McEvoy, T. J. 1998, *Legal Aspects of Owning and Managing Woodlands*, Island Press.

able during bug season. We know that relatively undisturbed, mature forests are important habitats for many of these species. Practicing ecosystem management in such a way as to protect nesting opportunities for **Neotropical** songbirds requires guidelines that will limit harvesting in certain forest types for a few weeks each year. This is a small price to pay to ensure the survival of these species, but some consider it an erosion of private property rights.

Managing whole ecosystems will require an unprecedented level of cooperation among neighboring woodland owners, municipalities, states, and countries. Uncertainty about how this cooperation will occur, especially at local levels, is a primary source of concern for owners, managers, and users of forests. Notwithstanding, successful implementation of ecosystem management worldwide will require drastic changes in forestry practice.

Appendix A
Index of Some Eastern Forest Types

Boreal Forest Region

Boreal conifers
Balsam fir
Black spruce
Black spruce–tamarack
White spruce
Tamarack

Boreal hardwoods
Aspen
Pin cherry

Northern Forest Region

Spruce-fir types
Red spruce
Red spruce–balsam fir
Red spruce–yellow birch
Red spruce–sugar maple–beech
Paper birch–red spruce–balsam fir
Northern white–cedar

Pine and hemlock types
Red pine
Eastern white pine
White pine–hemlock
Eastern hemlock
White pine–northern red oak–red maple
White pine–chestnut oak
Hemlock–yellow birch

Northern hardwoods
Sugar maple
Sugar maple–beech–yellow birch
Paper birch

Introduction to Forest Ecology and Silviculture

Black cherry–maple
Beech–sugar maple
Red maple

Other northern types
Gray birch–red maple
Black ash–American elm–red maple

Central Forest Region
White oak–black oak–northern red oak
Northern red oak
Black locust
Silver maple–American elm
Eastern red cedar

Source: Adapted from Eyre, 1980

Appendix B
Stocking Guides for Major Forest Types

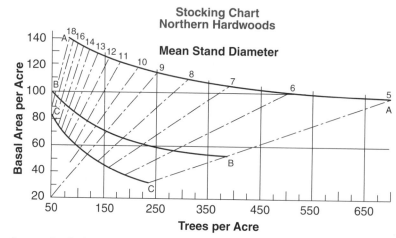

**Stocking Chart
Northern Hardwoods**

Source: Leak, Solomon, and Filip

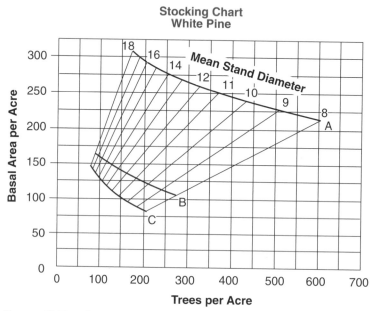

**Stocking Chart
White Pine**

Source: Philbrook, Barrett, and Leak

Glossary

Acid deposition — Formerly known as acid rain, it is the fallout of highly acidic compounds of nitrogen and sulphur that are pollutants resulting from industrial activities that use coal or other fossil fuels for energy. It is not called "acid rain" anymore because researchers have demonstrated that it does not have to rain for acids to be deposited.

Acre — An area of land containing 43,560 square feet, approximately equal to the playing area of a football field.

Advance reproduction — Well-established young trees that are capable of surviving after a regeneration treatment.

Aspect — The direction a slope faces.

Basal area — A measure of tree density. It is determined by estimating the total cross-sectional area of all trees measured at breast height (4.5 feet) and expressed in square feet per acre.

Biomass — The total weight of all harvestable vegetation from stands that is usually processed into chips. Biomass does not include the weight of the stump or roots.

Browse — Leaves, buds, and woody stems used as food by woodland mammals such as deer and moose.

Buffer strip — A protective strip of land or timber adjacent to an area requiring attention or protection; for example, a protective strip of unharvested timber along a stream.

Bull cruise — An estimate of standing timber volumes based upon a walk through an area designated for harvest, where the estimate is based solely on the visual observations of the person

---

Sources: Glossary definitions have been adapted from U.S. Forest Service-Green Mountain National Forest; McEvoy, 1992; Vermont Tree Farm Committee; McClain; and Trokey and Bergman.

doing the cruise without the aid of equipment or scientific procedures.

Bumpers — Trees near skid trails used as pivot points to turn a load of logs, usually resulting in severe injury to the stem of the bumper tree.

Canopy — The uppermost reaches of a more or less continuous cover of branches and foliage formed collectively by the crowns of adjacent trees and other woody growth.

Cleaning or weeding — Regulating the species composition of a young stand by eliminating some trees and encouraging others. It is usually noncommercial and involves freeing seedlings or saplings from competition with ground vegetation, vines, and shrubs.

Clear-cutting — A harvest that removes all trees from a designated area at one time, for the purpose of creating a new, even-aged stand.

Climax — An association of plants and animals that will prevail in the absence of disturbance; the culmination of a successional cycle. Also referred to as late-successional.

Clones — A collection of stems that have propagated by vegetative means from the roots of a single parent.

Crop trees — Trees that are selected in even-aged stands and favored because they are expected to grow for the entire rotation.

Crown — The aboveground portion of a tree extending up and out from the first main branches on the stem. (The root "crown" refers to the area where the tree trunk meets the ground.)

Cruise — A survey of forest stands to determine the number, size, and species of trees, as well as the terrain, soil condition, access, and any other factors relevant to forest management planning, primarily in anticipation of timber extraction.

Cull — Trees that have no current or potential market value for sawtimber.

Cutting cycle — The interval between harvesting operations when uneven-aged methods are employed using group or single-tree selection.

Cutting methods — Timber management practices employed to either regenerate a new stand (regeneration cutting) or to improve the composition and increase the growth rate of the residual trees in the existing forest (intermediate cutting).

Deferment cutting — A variation of the shelterwood method where the final removal is delayed at least until the first thinning of the new stand.

Density — The quantity of trees, basal area, volume, or some other measure, per unit of area.

Diameter or *d.b.h.* — "Diameter breast height" or "diameter at breast height," taken at 4.5 feet above average ground level around the base of the tree.

Dieback — Any condition where portions of a tree's crown dies due to conditions other than shading.

Disturbance — Any event that disrupts an otherwise stable pattern. The event can be abrupt, as in a fire, hurricane, or a timber harvest; or the event results in slow change, such as drought, disease or an insect depredation. The effect of disturbance is usually change in the species composition and/or structure of an ecosystem.

Dominant — A tree that is free grown under forest conditions so that at least a portion of its crown extends above those of surrounding trees.

Early-successional — Any association of plants and animals that are capable of surviving and reproducing under harsh environmental conditions. The first species to occupy an area after a major disturbance. Also known as "pioneers."

Ecosystem management — A new approach to managing forests that attempts to "meet human needs without disrupting the integrity of the ecological processes by which forest ecosystems sustain themselves." (David Brynn, Middlebury, Vermont, in a September 1995 presentation.)

Even-aged management — A timber management method that produces a forest or stand composed of trees having relatively small differences in age. The difference in age between trees forming the main canopy level of a stand usually does not exceed 20% of the age of the stand at rotation age.

Fallers — Those members of the logging crew whose job it is to fell trees so as to lessen stand damage and make it easy for others on the crew to extract the logs, usually with heavy equipment. Sometimes referred to as "choppers."

Feeder roots — The very finest portions of a plant's root system (usually within the top few inches of soil), which are primarily responsible for uptake of nutrients and moisture.

Final harvest — Under even-aged methods, the culmination of the current rotation when the overstory of the current stand is totally removed.

Forest health — A somewhat controversial term among foresters, ecologists, and others; as used here it implies that a healthy forest is one where the essential ecosystem integrity is maintained largely within the limits imposed by nature. The term also implies that careless or excessive human use can lead to a degradation of ecosystem integrity.

Forest regions — A classification of forest vegetation based mostly on climatic patterns that organizes common groups of tree species often found together over broad geographical areas. The groups are known as "forest types" or "forest cover types."

Forestry — The art and science of growing and managing forests and forest lands for the continuing use of their resources.

Forest type or forest cover type — A natural group or association of different species of trees that commonly occur together over a large area. Forest types are defined and named after one or more dominant species of trees in the type, such as the "spruce-fir" and the "birch-beech-maple" types.

Form — With reference to a tree, the degree of taper between diameter at the top of a 1-foot stump and diameter at the top of the first 16-foot log. Also known as "form class," especially when used in reference to volume calculations.

Free grown — A tree that has grown all of its life in forest conditions without being overtopped by other trees.

Grade — A measure (usually subjective) of a tree's ability to produce high-quality, defect-free lumber. Grading in logs and lumber is progressively less subjective.

Group selection — An uneven-aged regeneration method to remove small groups of trees to favor the reproduction and establishment of late-successional species.

Growing stock — All the trees growing in a stand; generally expressed in terms of number, basal area, or volume.

Habitat — The place where a plant or animal can live and maintain itself.

Hard mast — Tree fruits or seeds that are surrounded by a hard outer shell. Also popularly known as nuts. The fruits of oaks and hickories, for instance, are considered hard mast.

High-grading — An exploitive logging practice that removes only the best, most accessible, and most marketable trees in the stand.

High-risk — A tree or stand that will not survive another ten years, or will have a net loss of timber volume in the next ten years.

Improvement cutting — An intermediate treatment to improve the growth rate and vigor of residual trees.

Increment — The increase in diameter, basal area, height, volume, quality, or value of trees in a stand over time.

Intermediate treatments — In even-aged methods, any removal of trees from a stand, between the time of establishment and the final harvest, undertaken to improve residual stand growth and/ or to improve species composition or stand health.

Intolerant species — Plant species that are incapable of surviving in the shade of other trees.

Late-successional — (See "climax.")

Liquidation cutting — When the objective of logging is to remove all merchantable product from the forest with no regard to either stand improvement or regeneration. It often precedes an outright sale of the land.

Log landing — A cleared area near public roads, where logs are processed and loaded onto trucks.

Management — With reference to forests, control of assets to achieve a purpose. Good forest management is always controlled by objectives aimed at improving the forest conditions and values.

Maturity — In forest trees maturity is usually expressed in one of two ways: 1) financial maturity, when a tree has reached the point where it has maximized value growth from the perspective of the marketplace; and 2) biological maturity, when a tree has reached the point where the energy costs of maintaining itself exceed the energy input from photosynthesis. Most biologically mature trees are much older than the age of financial maturity.

MBF — One thousand board feet.

Mean stand diameter — Based on a relationship of the number of trees per acre and the sum of their basal area, it is the d.b.h. of the tree of average basal area.

Merchantable — A standing tree that has a net value when processed and delivered to a mill or some other market, after all harvesting and transportation costs are considered.

Mid-successional — An association of plant and animal species that replace early-successional species, but are eventually replaced by climax species in the absence of disturbance.

Mixed hardwoods — Timber stands characterized by a mixture of hardwood tree species, including oaks, basswood, white ash, hickories, soft maple, and others. A mixture of forest types.

Mortality — Trees that die before the end of the rotation and are usually not harvested.

Neotropical — Refers mostly to songbird species that migrate between temperate locations in North America, commonly for breeding, and tropical locations in Central or South America where they overwinter.

Northern hardwoods — Describes a collection of hardwood species composed mostly of sugar maple, yellow birch, beech, red maple and other associates that are collectively known as northern hardwoods. In this context it is best described as a "forest type." The "northern hardwood forest" refers to a major forest region in North America that includes these species and others.

Old growth — A self-perpetuating forest community that has reached a dynamic steady state (i.e., changes occur in the community only when gaps are formed as old trees die out, but the changes do not affect the overall character of the community) in the absence of silvicultural treatments. The dominant vegetation is considered to be climax with all age classes present.

Open grown — Refers to a tree that has grown most of its life without competition from surrounding trees.

Operable — In reference to harvesting, a set of conditions including site access, prescription difficulty, and potential product values that, in total, determine if a site is worth logging.

Overmature — A stand of trees that is older than normal rotation age for the type.

Overstory — The upper crown canopy of a forest, but usually stated in reference to the largest trees.

Pioneers — Shade intolerant species, such as aspen and paper birch, that are the first trees to invade recently disturbed sites and abandoned fields.

Pole timber — A d.b.h. size-class representing trees that are usually more than 4 inches d.b.h. but less than 10 inches. The term "pole-size" is commonly used interchangeably with pole-timber.

Precommercial thinning — An intermediate treatment in a young stand that does not yield any salable products.

Preparatory cut — The first phase of a three-cut shelterwood system that removes species that will not be favored in the next stand.

Prescription — A course of action to effect change in a forest stand. See also "treatment."

Productivity — Growth per unit time. In forestry, the number of growth rings per inch of diameter is often used as a measure of productivity.

Reforestation — The natural or artificial restocking of an area with trees.

Regeneration — The natural or artificial renewal of trees in a stand.

Regeneration cutting — Trees are removed from the stand to create conditions that will allow the forest to reproduce a new stand of trees. This is accomplished under either an even-aged management system or an uneven-aged management system.

Release — The freeing of well-established crop trees, usually large seedlings or saplings, from closely surrounding growth.

Removal cut — The final cut of the shelterwood system, which removes the remaining mature trees, releasing the young stand.

Residual — Trees that are left to grow in the stand following a silvicultural treatment.

Rotation — Usually considered to be the length of time it takes to grow an even-aged stand to the point of financial maturity.

Rotation age — The age at which a stand is considered ready for harvest under the adopted plan of management.

Salvage cutting — The removal of dead, dying, and damaged trees after a natural disaster to utilize the wood before it rots.

Sanitation cutting — The removal of dead, dying, and damaged trees after a natural disaster to prevent or interrupt the spread of insects or disease.

Sapling — Trees that are more than 4.5 feet tall but less than 4 inches d.b.h.

Sapwood — The first few inches of wood under the bark of a tree, usually lighter in color, composed of cells that mostly transport water and nutrients to the leaves.

Sawlog — A log considered suitable in size and quality for producing lumber. A standard sawlog is 16 feet long plus 4 inches for trim.

Sawtimber — Trees that have obtained a minimum diameter at breast height that can be felled and processed into sawlogs.

Scarification — Loosening the topsoil in open areas to prepare for regeneration by direct seeding or natural seed fall.

Seedlings — Trees that are less than 4.5 feet tall.

Seed tree — The removal of most of the trees in one cut, leaving a few scattered trees of desired species to serve as a source of seed for the new even-aged stand.

Selective harvest — A common term in forestry that has no silvicultural meaning. A selective harvest simply means that someone selected trees for harvest, but the criteria for selection are usually not specified.

Shade tolerance — The capacity of a tree to develop and grow in the shade of, and in competition with, other trees. Sugar maple is an example of a highly shade-tolerant species; American beech is another.

Shelterwood — A series of two or three treatments that gradually open the stand and stimulate natural reproduction of a new even-aged stand.

Silvics — Study of the life histories of tree species and the forest sites on which they evolved.

Silviculture — The art and science of growing forests for timber and other values.

Silvicultural — An adjective that implies silviculture is being used.

Single-tree selection — The removal of individual trees under uneven-aged regeneration methods. This term is sometimes confused with "selective harvest."

Site index — A measure of the relative productive capacity of an area, based on tree height growth.

Site preparation — An activity intended to make conditions favorable for planting or for the establishment of natural regeneration.

Skidder — A four-wheel-drive, tractor-like vehicle, articulated in the middle for maneuverability, with a cable or grapple on the back end for bringing logs from the stump to a landing area where they are loaded onto trucks.

Skid trail — Any path in the woods over which multiple loads of logs have been hauled. A trail that enters a main landing area is called a primary skid trail.

Slope — The average angle of incline of the terrain; usually expressed as a percentage based on the amount the incline rises over a horizontal distance.

Soil pH — A numerical designation of the acidity or alkalinity of a soil solution. Technically, pH is the common logarithm of the reciprocal of the hydrogen ion concentration of a solution. A pH of 7.0 indicates a neutral solution; higher values indicate increasing alkalinity, and lower values indicate increasing acidity. Most forest soils are slightly to highly acidic.

Stand — A community of trees occupying a specific area and sufficiently uniform in composition, age, arrangement, and condition as to be distinguishable from the forest on adjacent areas.

Standard — A tree of seed origin.

Stand condition — A silvicultural classification used to describe the present condition of a stand, particularly in relation to its need for treatment.

Stem — The aboveground portion of a tree between ground level and the first main branches in the canopy. Also referred to as "trunk" or "bole."

Stocking — An indication of the number of trees in a stand as compared to the optimum number of trees to achieve some management objective, usually improved growth rates or timber values.

Strip cut — A "clear-cut" laid out as a long, narrow strip.

Structure — The vertical and horizontal arrangement of vegetation. In silviculture, structure also refers to the relative ages of trees, as in "even-aged" or "uneven-aged," and diameter distributions. See also "stocking" and "density."

Stumpage — The value of timber as it stands in the woods just before harvest (on the stump).

Stump sprouts — Dormant buds under the bark of most hardwood species that begin to develop (usually at or just below the stump surface) into new stems when a tree has been cut or the stem is severely damaged.

Succession — The orderly and predictable replacement of one plant community by another over time in the absence of disturbance.

Sucker — A stem that forms from adventitious buds on the roots of some hardwood species, such as aspen and American beech, and that begins to develop when a tree has been cut or the stem and root system have been severely damaged.

Sustainability — The capacity of an ecosystem to provide benefits in perpetuity without substantially compromising ecosystem integrity.

Sustained yield — An annual or periodic output of products from the forest without impairment of the productivity of the land.

Thinning — Generally, a reduction in the number of trees in an immature forest stand to reduce tree density and concentrate growth potential on fewer, higher-quality trees.

Thinning shock — A slowing of tree growth brought on by acute exposure to higher temperatures and wind following a heavy thinning.

Timber stand improvement or *T.S.I.* — Activities conducted in young stands of timber to improve growth rate and form of the remaining trees.

Tolerance — The ability of a tree to grow satisfactorily in the shade of, or in competition with, other trees. Trees that are classified as tolerant can survive and grow under continuous shade.

Treatment — Any action in forest stands that is controlled by a silvicultural prescription.

Understory — In forests, the vegetation that occupies an area between the forest floor and the main canopy of the stand.

Uneven-aged or *all-aged management* — A timber management method that produces a stand composed of a wide range of ages and sizes.

Volume table — A table that utilizes d.b.h. or log diameters and log heights (usually 16 feet) or actual log lengths to estimate board foot volumes according to a set of assumptions (known as the "log rule") about how the log will be processed into boards. A commonly used log rule in the region is the "International ¼-inch rule."

Well-stocked — Referring to stands having sufficient growing stock for sawtimber production.

Whorl — In conifers, a group of branches that emerge from the stem in a single growing season all at the same height, separated by clear stem for a foot or two until the next year's whorl.

Wildlife habitat — The sum total of environmental conditions of a specific place occupied by a wildlife species or a combination of such species. The three main elements of habitat are food, water, and cover.

Windthrow — A forest tree that has been toppled due to high winds, usually coupled with wet ground conditions. It is also a common phenomenon along the margins of strip cuts and clear-cuts.

Yard — Any place where animals congregate for protection from the elements. In extreme northern portions of the northeast region, deeryards are associated with coniferous cover.

Yield — Total forest growth over a specified period of time, less mortality and unmarketable fiber and cull.

Yield table — A species-specific representation of the amount of usable wood fiber a forest site can be expected to produce during a single rotation, based upon site index.

References
and Further Reading

References

Beattie, M., C. Thompson, and L. Levine. *Working with Your Woodland*. Hanover: University Press of New England. 1983.

Bergdahl, D. Personal communication. University of Vermont, Burlington. 1998.

Daniels, T., J. A. Helms, and F. S. Baker. *Principles of Silviculture*. Second Edition. New York, New York: McGraw Hill. 1979.

Eyre, F. H. *Forest Cover Types of the U.S. and Canada*. Society of American Foresters. 1980.

Fowells, H. A., editor. *Silvics of Forest Trees of the United States*. U.S. Department of Agriculture. Agricultural Handbook No. 271. Washington, DC: U.S. Forest Service. 1965.

Leak, W., and J. Riddle. *Why Trees Grow Where They Do in New Hampshire Forests*. USDA Forest Service, Northeastern Forest Experiment Station. NE-INF-37-79. 1979.

Leak, W. B., D. S. Solomon, and S. Filip. "A Silvicultural Guide for Northern Hardwoods in the Northeast." USDA Forest Service Research Paper NE-143. 1969.

Leopold, A. *A Sand County Almanac*. New York, New York: Ballantine Books. 1949.

McClain, J. Appendix to client's management plans. 1993.

McEvoy, T. J. *Using Fertilizers in the Culture of Christmas Trees* (First Edition). Hinesburg, Vermont: Paragon Books. 1992.

McEvoy, T. J. *Legal Aspects of Owning and Managing Woodlands*. Washington, D.C.: Island Press. 1998.

Philbrook, J., J. P. Barrett, and W. B. Leak. "A Stocking Guide for Eastern White Pine." USDA Forest Service Research Note NE-168. 1973.

Pritchett, W. L., and R. F. Fisher. *Properties and Management of Forest Soils.* Second Edition. New York, New York: John Wiley and Sons, Inc. 1987.

Sendak, P., N. K. Huyler, and L. D. Garrett. "Lumber Value Loss Associated with Tapping Sugar Maples for Sap Production." Northeastern Forest Experiment Station Research Note NE-306. Burlington, Vermont: USDA Forest Service. 1982.

Seymour, F. C. *The Flora of Vermont.* Burlington: Vermont Agricultural Experiment Station, University of Vermont. 1969.

Smith, D. M. *The Practice of Silviculture,* Eighth Edition. New York, New York: John Wiley and Sons, Inc. 1986.

Trokey, C. B., and F. Bergman. "Forestry Terms for the Woodland Owner." University of Missouri Extension. 1990.

U.S. Forest Service-Green Mountain National Forest. *Developing Markets for Low Quality Wood in Bennington County.* Glossary. Rutland, Vermont: USDA Forest Service. 1990.

Vermont Tree Farm Committee. "Definitions of Commonly Used Forestry Terms." Burlington, Vermont. 1983.

Wegner, K. F. , editor. *Forestry Handbook.* Second Edition. New York, New York: John Wiley and Sons, Inc. 1984

Further Reading

Daniels, T. W., J. A. Helms, and F. S. Baker. *Principles of Silviculture.* Second Edition. New York: McGraw-Hill. 1979.

Decker, D. J., et al. *Wildlife and Timber from Private Lands: A Landowner's Guide to Planning.* Ithaca, New York: Cornell Cooperative Extension Service. 1990.

DeGraaf, R. M., and D. D. Rudis. *New England Wildlife: Habitat, Natural History and Distribution.* USDA For. Serv. Gen. Tech. Rep. NE-108. 491 p. (Source: USFS, Durham, NH: 603-868-7690). 1986.

Jones, G. T. 1993. *A Guide to Logging Aesthetics: Practical Tips for Loggers, Foresters, and Landowners,* NRAES–60. Ithaca, New York: Natural Resource, Agriculture, and Engineering Service. 1993.

Tubbs, Carl H., et al. *Guide to Wildlife Tree Management in New England Northern Hardwoods.* USDA Forest Service NAFES Gen. Tech. Rep. NE-118. 1986.

Other Publications from NRAES

Note: The publications listed below are just a few of the publications from NRAES. Over 160 publications are available; call for a free catalog.

The Careful Timber Harvest: A Video Guide to Logging Aesthetics
SPNHF–1 • 20 minutes long • This video tells what loggers think about logging aesthetics and teaches the viewer practical, step-by-step techniques proven to address concerns during and immediately following harvesting. Foresters will learn how to show clients methods to reduce impacts of logging and how federal cost-sharing programs can help reduce costs. Loggers will learn how to minimize disruptive impacts and understand the importance of aesthetics. And landowners will learn to better visualize the steps in a timber harvest and discover what areas of a harvest have the greatest potential impact on aesthetics. (1994)

Enhancing Wildlife Habitats: A Practical Guide for Forest Landowners
NRAES–64 • 172 pages • This publication contains recommendations and field exercises for landowners who want to ensure quality habitats for wildlife in the region. It includes general sections on forest ecology, understanding wildlife habitats, and wetlands, as well as specific guidelines for enhancing habitats of woodcock, ruffed grouse, white-tailed deer, wild turkey, and other upland and wetland animals. The book contains over 100 figures, many of which are black-and-white photographs; 11 tables; and a glossary. (1993)

A Guide to Logging Aesthetics: Practical Tips for Loggers, Foresters, and Landowners
NRAES–60 • 28 pages • Today's public and private forestland owners are placing greater emphasis on ecological, recreational, and aesthetic values than on timber production. This publication addresses planning and conducting a timber harvest to mini-

mize the disruptive effects of cutting and removing trees on a forest. Topics covered include concerns, solutions, truck roads, landings, skid trails, tree felling, administration and planning, and costs. Also included are 50 color photos, a glossary, references, and a list of state agencies. (1993)

Lumber from Local Woodlots

NRAES–27 • 42 pages • This guide provides background knowledge of the woodlot-to-lumber process. Topics covered include wood species, wood properties, sources of professional assistance and training, proper woodlot management, contracting with loggers and sawyers, good harvesting practices, sawing methods, and lumber drying and storage. This guide is meant to encourage the use of local woodlot resources for construction. (1988)

Using Fertilizers in the Culture of Christmas Trees

NRAES-127 • 179 pages • This book is a factual, comprehensive reference geared specifically toward Christmas tree plantations. It aims to help growers understand the principles of fertilization and apply this knowledge to growing healthier, higher quality trees in less time. Chapters cover cultural objectives, theories of plant nutrition, understanding soils, nutrient uptake, analyzing the nutrient status of the plantation, fertilizer materials, placement and timing of applications, potential fertilizer responses for important species, and other topics. Twenty-six figures, 14 tables, a glossary, an index, and two appendixes are included. (2000 Revision)

Ordering Information

Publications listed on pages 87–88 can be ordered from NRAES. Before ordering, contact NRAES for current prices and for exact shipping and handling charges, or ask for a free copy of our publications catalog. See the inside back cover for more information about NRAES, including a list of member universities.

Natural Resource, Agriculture, and Engineering Service (NRAES)
Cooperative Extension • 152 Riley-Robb Hall
Ithaca, New York 14853-5701

Phone: (607) 255-7654 • Fax: (607) 254-8770
E-mail: NRAES@CORNELL.EDU • Web site: WWW.NRAES.ORG